Classroom Close-Ups: 4

Mathematics: Friend or Foe?

Classroom Close-Ups

A series edited by Gerald Haigh which tries to answer the questions 'what really happens?', and 'what really matters?', in education.

Titles published are:
1 *'Integrate!'* by Gerald Haigh
2 *Teaching Morality and Religion* by Alan Harris
3 *Early Reading and Writing* by Ramin Minovi
4 *Mathematics: Friend or Foe?* by Dorothy Evans
5 *Education for Sale* by Eric Midwinter

and forthcoming books will cover:
Pre-School Education
Teaching Art

Classroom Close-Ups: 4

Series Editor: Gerald Haigh

MATHEMATICS: FRIEND OR FOE?

A review of some ideas about learning and teaching mathematics in the primary school

Dorothy Evans

London
GEORGE ALLEN & UNWIN
Boston　　　　Sydney

First published in 1977

This book is copyright under the Berne Convention.
All rights are reserved. Apart from any fair dealing for the
purpose of private study, research, criticism or review,
as permitted under the Copyright Act, 1956, no part of
this publication may be reproduced, stored in a retrieval system,
or transmitted, in any form or by any means, electronic,
electrical, chemical, mechanical, optical, photocopying,
recording or otherwise, without the prior permission of the
copyright owner. Enquiries should be addressed to the
publishers.

© George Allen & Unwin (Publishers) Ltd, 1977

ISBN 0 04 372022 6 hardback
     0 04 372023 4 paperback

Printed in Great Britain
in 10 on 11 point Press Roman
at the Alden Press, Oxford

The ceaseless round, the common task...

Mathematics is everywhere, the universe, the sun, the moon and the stars.
Mathematics tells us where we are and where we are going to.
The distance from here to there – and back again.
Mathematics is our income tax, our salaries, our homes and our way of life.
Mathematics is the science of number and shape and of the link between them.
It is a language we learn so that we can count our sheep, our money and divide our fellows into different social classes.
A language which enumerates the ten commandments, the seven wonders of the world and the three blind mice.
Mathematics is the balance of payments and the imbalance of the world's supply of food.
Mathematics tells us that one with one make one, but also that one and one make two.
Where numbers are integral, rational, complex or entirely imaginary.
Mathematics is the logical development of fundamental abstract ideas, given substance through words, numbers and symbols.
Developed through the ages, since time began, by Archimedes, Thales, Euclid, Eratosthenes, Pythagoras and other giants, on whose shoulders, like Newton, we may stand.
Mathematics is in the force of the tempest, the rain, the rainbow and the heat of the sun.
It measures the water in the sea, the sand in the desert and the height of the highest mountain, and the men who risk death to conquer them.
Mathematics is the air we breathe, the space we live in, light and darkness, sunshine and shadows.
When Winter comes, darkness lengthens but the rotating world pursues its elliptical path to bring us once more to Spring and a new beginning.

J. L. PITTS

Acknowledgements

One of the advantages of working in the mathematics department of a college of education is that one is constantly meeting new enthusiasts in the subject. Colleagues, students, local teachers and advisers are keen to talk about mathematics, to exchange ideas and to share problems. During the last few years I have stored away comments, stories, examples and ideas and have now been tempted to write some of them down in this small book. In doing so, I would like to acknowledge the enthusiasm and the creative skill of the people with whom I have been working, whose opinions are reflected here.

In particular, I would like to thank Helen Ball, Michael Foot, Jean Gardiner, Elizabeth Page, John Porter, Moira Proudfoot and Blake Stimpson for discussing at length with me their work and experiences, Frank Carter for his helpful comments, his time and his mathematical advice and Josie Pitts for her skilful typing, her encouragement and her poetic efforts.

I acknowledge also, of course, the main sources used in my brief survey of some mathematical schemes (listed on page 63) and particularly thank Mrs J. Lister for allowing me to use *Guidelines in School Mathematics* (prepared by the Mathematics Department of the Manchester College of Education).

F.I.D.E.
January, 1976

Contents

Introduction: 'Red, Amber, Green'   page   13

CHAPTER 1: What Some People Say About Learning and Teaching Mathematics

1.1	Numerical Competence	17
1.2	Utilitarian Purposes	19
1.3	A Rag-bag of Techniques	21
1.4	Practical Skills	25
1.5	The Physical World	26
1.6	A Fascinating Study	27
1.7	Technological Advance	32
1.8	A Means of Communication	35
1.9	A Form of Beauty	38
1.10	Order out of Chaos	43
1.11	Logical Thinking	45
1.12	Accuracy	52
1.13	Ingenuity and Resourcefulness	54
1.14	An Explanation of Natural Phenomena	55
1.15	A Study of Sets	57

CHAPTER 2: The Syllabus

2.1	An Ad Hoc or Structured Approach	61
2.2	Schemes of Work to be Analysed	63
2.3	Plan of the Analysis	64
2.4	The Pre-Number Stage to Awareness of Numbers to 100	65
2.5	The Four Operations on Number	70
2.6	Fractions, Decimal Fractions and Percentages	75
2.7	Measurement	80
2.7.1	Length, weight (mass) and capacity	80
2.7.2	Length, area and volume	84
2.7.3	Volume and capacity	87
2.7.4	Angles	89
2.7.5	Time	91
2.7.6	Money	92
2.8	Shapes	93

2.9	*Graphical Representation*	95

CHAPTER 3: The Changing Scene

3.1	*Recent Developments*	101
3.2	*Innovation*	102
3.2.1	*Reasons for change*	103
3.2.2	*The fact of change*	106
3.3	*The Attributes of Change*	107
3.3.1	*The child*	107
3.3.2	*The range of work*	107
3.3.3	*Resources: textbooks, structural apparatus, help for the teacher, equipment*	108
3.4	*The Structure Within the Class, School and Neighbourhood*	115
3.5	*The Vehicles of Change*	116
3.5.1	*The individual*	116
3.5.2	*Physical resources*	117
3.5.3	*Organisational structure of the school*	119
3.6	*The Effects of Change*	122
3.6.1	*General effects*	122
3.6.2	*Specific effects: the teacher, the child, authors of textbooks, teacher training, mathematical content, methods of analysis and recording in the classroom, parents*	123
3.7	*The Verdict So Far?*	130
3.8	*Afterthought*	130
Appendix: Testing		132
Bibliography		136
Index		138

Introduction: 'Red, Amber, Green'

Like the ostrich, some of us who teach mathematics hide our heads in the sands of uncertainty and half-thought-out principles, trying to believe that what we are actually doing in our own classrooms is the best for the children and for ourselves.

The aim of this book is to look critically at some of the current practices and theories related to the teaching of mathematics. This may then help to clarify our own views about the purposes of learning and teaching mathematics, about the content of the mathematics introduced into the classroom, about the reasons for making this particular selection from the vast network of mathematical knowledge, about the methods being used to teach mathematics and the justification made for these methods. In short, this book seeks to explore the 'why', the 'what' and the 'how' of bringing mathematics into school. This will be done by looking first at the varying claims made by the teachers of mathematics for its inclusion and its importance in the curriculum, then by discussing the content of a range of commonly used syllabuses and finally by considering a number of methods and approaches currently being used in schools.

I do not expect the reader to accept the views which are expressed here, nor is it desirable that he should. My purpose will be served if we are all caused to stop and think again what we are doing under the guise of teaching mathematics, and whether we are convinced that what we are doing cannot be bettered. Such a continuous process of self-examination requires from the teacher not only knowledge, but flexibility, adaptability and no small measure of courage!

Many claims have been made by teachers of mathematics about their work. A number of these claims have been selected for discussion here. The reader may recognise and welcome some of them and be tempted to settle into complacency, but, while every one of these statements is true to a greater or lesser extent, it is important that they are put into perspective in the overall picture of school mathematics. There is at present much uncertainty and confusion, causing many of us to feel vaguely uneasy about our teaching of mathematics, wondering whether it is still respectable, feeling convinced that in some magical way everyone else knows a lot more than we do ourselves but fearing to show our ignorance of what is really going on.

So, adopting the analogy of the familiar traffic lights, I believe that it is not only desirable but essential to recognise and to take advantage of the red light in whatever form it comes, to give ourselves the period of reflection and preparation afforded by the amber light, in order to proceed with a certain amount of confidence and pleasure at the green light.

It is essential that the teacher of primary school mathematics gets clear in his own mind:

(a) what his children need to know,
(b) how he can best use his own skill, expertise and knowledge to impart this knowledge, and
(c) what organisation best suits the particular children in his class such that their learning is maximised.

The immensity of the range of mathematical activities which have recently, for better or for worse, become part of the work done in primary schools has brought with it a disastrous tendency for children to spend a little time on a lot of things. It is essential that, at the beginning stages of acquiring a new concept or learning a new skill or technique, the child should have ample opportunity to practise and to consolidate and to apply his new knowledge before being whisked off to some other activity. It is far more satisfactory if, at this early stage, the child has a chance of mastering the concept rather than finding out later on that he has got to go back to relearn something which he has met before. It is the experience of many teachers that to motivate the child in the secondary school to relearn basic mathematical facts which were first introduced in the primary school is a very difficult task. It is also the experience of teachers that there is a stage in the acquisition of a new concept or skill when the time is ripe for a hard period of learning and consolidation to take place. If this optimum time is allowed to pass, the best opportunity for establishing the concept will have been lost. Throughout the primary school years there are many such optimum times and the teacher must recognise that frequently it is necessary for children to do work which is a grind and repetitive. This may not be popular. I agree wholeheartedly with the comment of a fellow teacher

'It is no preparation for life to condition children to do only that which is enjoyable and interesting and to skate round the rest in the pious hope that this also will be added unto them. This does not produce children who enjoy mathematics. It produces children who are only prepared to do mathematics which they enjoy.'

Therefore, bearing this in mind, let us examine critically some of the claims made by some people about learning and teaching mathematics, then decide which of the acceptable ones have priority and which to reject altogether.

Chapter 1

What Some People Say About Learning and Teaching Mathematics

1.1 NUMERICAL COMPETENCE

SOME PEOPLE SAY THAT:

'To develop numerical competence in a child is the most important aim of the teaching of mathematics.'

This is a familiar and commonly held opinion. It is absolutely true that a real understanding of numbers is vital. Numbers are a part of our language. We need to be able to recognise them, to use them, to combine them using the four operations, to know their quantity, their make-up, their relationship to their fellows, their uses, their accuracy in context, their limitations and possibilities, their size in relation to the context, their orderliness and pattern.

For instance, if we pick out any number at random we can claim to know that number if we can quickly call to mind several facts about that number. As adults we know that the more we use a concept the more familiar it becomes. With very little difficulty we might produce a list like that in Table 1.1.

A young child's story of a number might not include all the operations involved in the list in Table 1.1 but nevertheless the facility to recall a number of facts appropriate to the age and situation is a useful asset.

Is 27 a big number? However much we know a number we are not able to answer this question unless we understand it in its context. Statements like the following ones give a very different impression from each other of the size of 27:

The town raised £27 to send to Oxfam.
The school raised £27 to send to Oxfam.
The car skidded for 27 metres.

18 Mathematics: Friend or Foe?

The car was driven forward for 27 metres.
27 children were absent with measles.
27 children were absent with colds.

To take our place in society, we certainly need to appreciate numbers of all sizes. Almost every newspaper article makes mention of numbers, and the ways in which numbers occur in the text often cause us to use the four basic operations, namely, addition, subtraction, multiplication and division. However, when did we last need to divide

Table 1.1   '27'

$20 + 7$	$(13 \times 2) + 1$
$7 + 20$	Not prime
$10 + 10 + 7$	½ of 54
$(2 \times 10) + 7$	Less than 30
$30 - 3$	$3(10 - 1)$
3×9	3^3
9 N	
9×3	$2 + 3 + 4 + 5 + 6 + 7$ (Why?)
$(5 \times 5) + 2$	$10 + 9 + 8$
$5^2 + 2$	$11 + 9 + 7$
$3(4 + 5)$	Odd
Divisible by 3	

475 by 26, or multiply 289 by 71? Certainly we have occasion to require an answer to the combination of a couple of two-digit numbers frequently in our day-to-day negotiations with each other. Would not a thorough understanding of these simpler numbers and their relationship to each other give a better basis on which to build more advanced mathematical concepts, rather than a memorisation of the more involved techniques we tend to employ when dealing with larger numbers? For instance, suppose we have the problem 'what do fifteen thirty-fives make?' In view of the fact that we know all about 15 and 35, we can break down our task to be, say:

   (10×35) and (5×35)
or (30×15) and (5×15)
or (30×10) and (30×5) and (5×10) and (5×5)
or (10×30) and (10×5) and (5×30) and (5×5)
or $15 + 15 + 15 + 15 + \ldots$(thirty-five times!)
or $35 + 35 + 35 + 35 + \ldots$(fifteen times)
or ...

So, surely now that we know there are many ways of solving this problem, we should have confidence to find a method of tackling a more complicated multiplication sum. Being given sufficient opportunity to think around manipulations of this kind is likely to give rise to a desire to streamline the whole procedure and to economise on effort; hence a 'system' or technique is sought for, based more securely on one's own experience than on 'cashing-in' on memorised techniques which were the fruits of someone else's experience. As a teacher, I know that many children cannot be expected to derive techniques for themselves unaided, and many of the techniques they do derive will be unorthodox and unwieldy, but at least they will be more prepared to transform their own inventions into a recognised form.

It is likely that we agree that it is just as important as it has ever been that we are familiar with the basic number bonds of addition and multiplication – at least as far as the nine times table.

A secondary school teacher has written to me complaining bitterly of the dangers of creating 'a further wave of junior school teachers who dabble in all aspects of mathematics, whilst falling down on the basic job of giving a grounding first and foremost on handling numbers consistently and competently, which would give secondary specialists a fighting chance of actually teaching mathematics'. He is not alone with his criticism and there is much truth in it. The problem of helping a child to acquire numerical competence is one which has been taxing teachers for several generations. Just as there are undoubtedly many children now who are regrettably numerically inadequate when they leave the primary school, so there are also many adults, who have had various forms of teaching and experience, who also are numerically inadequate. So the problem is still with us and we need to re-examine our methods so that no opportunity is lost to promote numerical competence.

Let us take a critical look at the content, context and extent of the number work we include in our teaching and make sure that the emphasis we place upon it is adequate and appropriate. We are the inheritors of a powerful, sophisticated number system, within which it is possible to perform elaborate calculations with comparative ease. What a pity it is that so many people fail to realise the beauty and simplicity of the tool in their hands!

1.2 UTILITARIAN PURPOSES

SOME PEOPLE SAY THAT:

'Unless the mathematics we teach is useful, it is a waste of time.'

A postgraduate sociology student once said in discussion that the only useless subject he studied at school was mathematics. This remark gave me considerable food for thought. Putting aside his possible prejudices and preferences, I wonder what emphases were made in the mathematics teaching at his school? What did he regard to be mathematics anyway? How is the word 'useless' to be defined? Perhaps he was recalling a course which had been too theoretical (or too practical), or in which there had been an imbalance of some sort. But how unfortunate, even if this is the case, that the opportunity to consider some of the most sophisticated yet simple solutions created by man to deal with his problems should ever be presented to a child in such a way that permanent alienation is caused.

'Useful' mathematics would include, presumably, all forms of numerical and physical measurement. These would include counting, numerical calculations, length, weight, capacity activities, measurements in space, and the passage of time. These do, of course, form a vital part of the body of knowledge with which we should be equipped and we are failing to fulfil our mandate to society if the children we teach do not become familiar with them. The difficulty is that it is impossible to arrive at a consensus of opinion among a group of teachers about what the extent of this necessary and desirable utilitarian knowledge is.

In general, teachers of very young children have achieved a good balance because it is what is happening *now* which is important at that age, so the question of introducing things which the child will need 'later on' does not arise. But when should we start looking to the future? Each child normally arrives at the stage when he can predict the possible outcome of a line of action: 'If I do this...then that will happen.' This may be our cue as teachers to start to look to the future needs of the child; but what exactly does this imply? Do we teach 8-year-olds about buying expensive articles, 12-year-olds about hire-purchase and 14-year-olds about stocks and shares? There is a great danger that adults will superimpose adult problems on the child's experience, forgetting that the child's *own* problems are important to him. Mathematics supplies us with tools for helping to solve problems, but let us endeavour to put the appropriate tool in the right hands at a suitable time. The guidelines could be:

(i) Is the user of the tool capable of using it with understanding?
(ii) Has he sufficient confidence in his own powers to be willing to tackle the problem?
(iii) Does he know enough about the range of tools available to enable him to select the best tool for the purpose?

Suppose, for example, the problem facing an 8-year-old is that he wants a square of cardboard of a certain size in order to complete a model. To start with he has three possible lines of action:

(a) Use a ready-made one (in which case his problem wasn't really a problem to solve at all).
(b) Ask someone else to provide him with one.
(c) Set about making one for himself.

As teachers, I have no doubt that, failing line of action (a), we would prefer that (c) should take place. An analysis of the steps the child might take in solving this problem could be those suggested in Figure 1.1.

So rather than asking a child to solve problems based on an adult world, using methods of solution which adults have devised already, the teacher would do well to encourage children to identify problems in their own sphere and to help them to develop strategies based on a logical process of thought (see page 22) which they are likely to be able to use in any situation, by recalling past experience and knowledge of similar situations and by using available resources. Our child with the problem of producing the cardboard square will thus be able to devise a plan of campaign which might involve accumulating some specific pieces (scissors, card, pencil, template), recalling some past experiences and skills (drawing, measuring, cutting, knowledge of the basic properties of a square) and putting into action his plan, based upon a minimal but necessary amount of preparatory thought.

The utilitarian aspect of mathematics must take its important place in the school curriculum. It is necessary that children begin to acquire a 'tool box', but it is equally necessary that the tools from the box are well understood and that with their use the child may become aware of opportunities to extend his knowledge and powers of imagination and ingenuity.

1.3   A RAG-BAG OF TECHNIQUES

SOME PEOPLE SAY THAT:

> 'Mathematics is nothing but a rag-bag of techniques for doing this and doing that.'

At first sight this appears to be a most derogatory statement about mathematics but it must be borne in mind that rag-bags can be extraordinarily useful provided that one can find the piece one wants – or the next best thing! This raises several important issues. The first concerns memory. No one will deny that having a good memory is a

22 *Mathematics: Friend or Foe?*

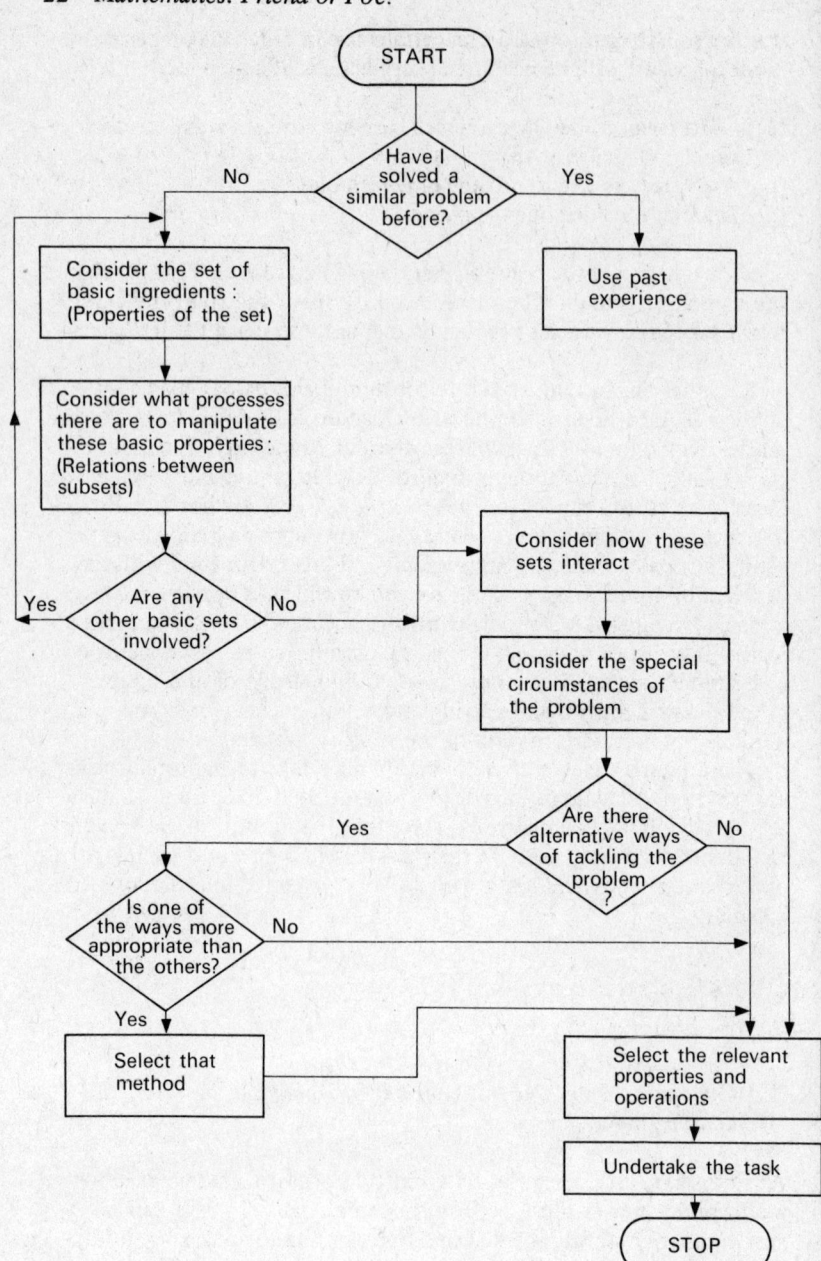

Figure 1.1 *There is a problem to solve.*

What Some People Say 23

desirable attribute not only from the point of view of becoming fully aware, knowledgeable members of society, but also because it ensures that problems encountered will be solved (if there is a solution) in the most economical way. Some people maintain that there is no need to store tables facts in one's head provided that one can reason them out, but what an extraordinarily uneconomic, inefficient way of arriving at, say, 9 × 7, when to recall the answer immediately is so much better. And how stupid it would be (and perhaps impossibly difficult!) to have to reason out the area of, say, a circle from first principles if one can recall instead a simple formula which one understands.

The study of mathematics provides an excellent arena for memory training – that is, training for a purpose – so perhaps we had better not belittle the value of the rag-bag of techniques from which we can extract the desired piece! But this analogy raises a second issue, which is that of needing a quick and reliable retrieval system so that required facts can be quickly sifted out. This necessitates developing a systematic way to do this and, once again, we are back at the need for a basic problem solving strategy.

Every one of us has a collection of techniques which we can recall with varying degrees of success and for which we do not bother to recall the mathematical backing when we are using it. For instance, I do not suppose many of us consciously bother to think why we just 'add a nought' (what an ambiguous phrase!) when we wish to multiply a

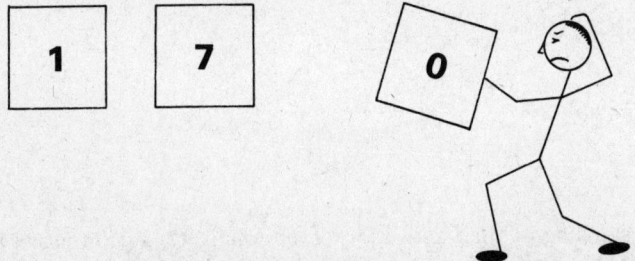

number by ten. And why, when adding fractions, do we use a common denominator, whereas when multiplying fractions we do not?

$$\frac{7}{12} + \frac{8}{12}$$

$$= \frac{7 + 8}{12}$$

$$= \frac{15}{12}$$

$$\frac{7}{12} \times \frac{8}{12}$$

$$= \frac{56}{144}$$

24 *Mathematics: Friend or Foe?*

Both of these processes are unquestioned by us because we know that at some time in the past we looked at them from first principles. We accept and use such simple formulae without continually testing their truth. Similarly, we have adopted various computational rules of thumb. All of these are acceptable practices for those who have an

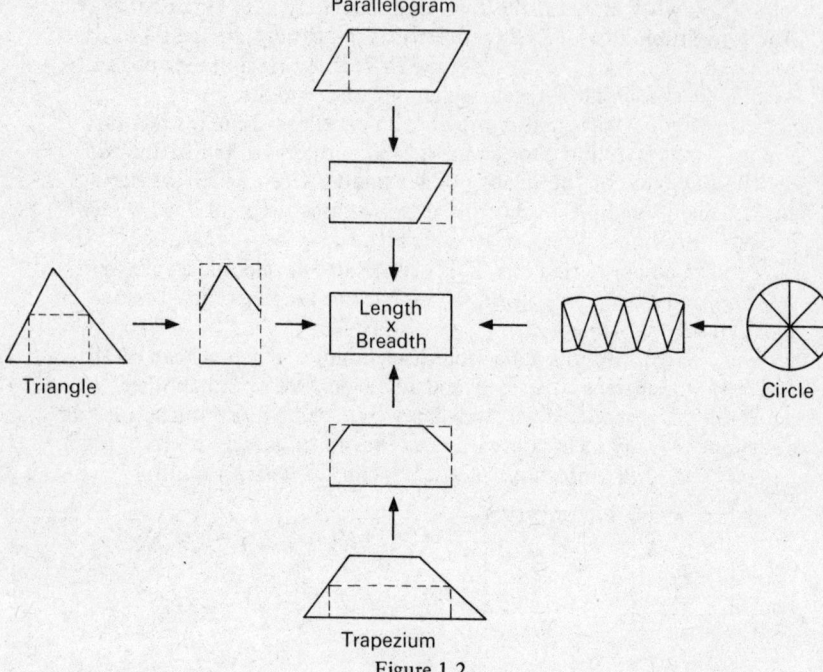

Figure 1.2

understanding of their derivation or, alternatively, a good memory, but we shall only succeed in baffling and alienating children if we try to impose our well-used techniques on them.

A third issue which arises from this analogy of the rag-bag is that occasionally the exact piece we want is nonexistent, and this provides the occasion when we need to use our ingenuity to find a good substitute. How often do we come across a problem, the exact likeness of which we have not met before, but for which our rag-bag can at least produce some helpful starting points.

It does not matter, for instance, that a search through the rag-bag fails to reveal the means for finding the area of a triangle, provided that the searcher can appreciate the relationship between the area of the triangle and the area of the rectangle in which it would stand. In fact,

What Some People Say

the *only* area formula that we need to store away is that of a rectangle, because the areas of all other shapes, such as triangles, parallelograms, trapezia and even circles, can easily be derived by (in imagination or otherwise) cutting up the shape and rearranging the pieces to form a rectangle (see Figure 1.2). As mentioned before, though, this is a waste of time if the correct formula is readily to hand.

To test out this principle of using a substitute piece of knowledge, give yourself the task of solving a problem the process for which you have either forgotten or you never knew anyway. Possibly a suitable example might be: what is the size of the interior angle of a regular pentagon?

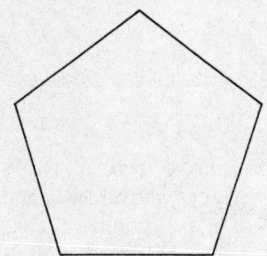 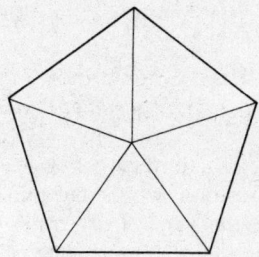

Your thoughts may have followed some line similar to this:

The interior angle of the pentagon is certainly bigger than that of a square which is 90 degrees.
The angles of a triangle add up to 180 degrees.
If, from some point inside the pentagon, spokes are drawn out to each of the vertices, five triangles will appear.
In each triangle the angles add up to 180 degrees.
At the centre of the spokes, i.e. the hub, the angles meeting there add up to 360 degrees.
So all the angles near the edge of the figure add up to (5 × 180)−360.
So each of the angles must be that result divided by 5 ...

True, this piece of reasoning would have been unnecessary if the formula $(2n - 4)/n$ right angles had been at hand, but how many of us are likely to store away such 'useless' pieces of information?

1.4 PRACTICAL SKILLS

SOME PEOPLE SAY THAT:

> *'Mathematics gives us the opportunity to teach some practical skills.'*

Which practical skills are referred to here? No doubt such skills as using measuring instruments correctly, using drawing instruments, knowing which instruments to use in a given situation, making diagrams, recording data graphically and economically, making models, tabulating data, using ready reckoners, using calculating devices.... There would seem to be little doubt about the truth of this claim for the inclusion of mathematics in the curriculum but these skills are but a few of the pebbles on the vast beach of mathematics. This beach is possibly more truly represented by the next claim.

1.5 THE PHYSICAL WORLD

SOME PEOPLE SAY THAT:

> *'Mathematics provides us with a way of looking at the physical world and getting to understand what we see.'*

We see a universe - or, at least, parts of it. It consists of people, places and things, words and numbers and sounds, shapes, movements and pattern...and many other things! Because of man's comparative immobility, his main universe of discourse is his immediate surroundings in which he familiarises himself with those sets of people, places, objects, shapes and numbers which have a direct bearing on his livelihood. He learns more and more about the properties of these sets, he classifies them (see page 43) according to a labelling system which he invents for himself or which he has devised with his fellow men. He looks for and uses relationships between these classified subsets. And what is mathematics but the study of these relationships? I hope to show a little later on (page 57) that any item included in primary school mathematics can be expressed as a relationship between one, two or more of the sets of peoples, places, objects, shapes and measures, and numbers. By thinking in terms of sets and their properties and relationships, we have a way of looking at our physical world with all its isolated elements, disunity and diversity and, from this study, detecting order, coherence, purpose, reason and beauty.

Take, for instance, the study of these three shapes:

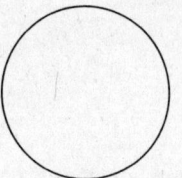 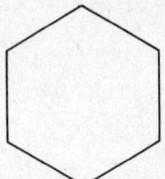 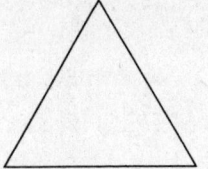

On first glance we would probably decide that they are different from each other. But let us now find in what ways they would be classified as being the 'same' as each other:

(i) They all enclose a space.
(ii) They are polygons.
(iii) They are regular shapes.
(iv) They are printed on the page (this is not such a trivial description as it may appear to be!)
(v) Each has a continuous border.
(vi) Each has bilateral symmetry.
(vii) Each has more than one axis of symmetry.
(viii) Each has rotational symmetry.
(ix) They are two-dimensional shapes.
(x) They are natural (as opposed to man-made) shapes.
(xi) They are topologically equivalent to each other.
And so on.

In our search for these 'samenesses' we have been caused to study this simple set of shapes very closely and hence have come to know them well. This will mean that on the next occasion when these particular shapes are involved, we shall know them a little better than we did before and so will be able to develop our knowledge into a more sophisticated form. And so learning takes place.

1.6 A FASCINATING STUDY

SOME PEOPLE SAY THAT:

> 'Mathematics is a fascinating study in its own right. It does not need to be useful to justify its existence.'

Whatever aspect of mathematics is being considered, there is no doubt that there is always something intriguing or intensely satisfying to the person who is intent on exploring a little beyond the situation itself. Take numbers, for instance – say, the sequence of triangular numbers

$$1, 3, 6, 10, 15, 21 \ldots$$

We don't need to think very deeply to decide which numbers come next, so instead we will set ourselves the task of listing as many clues as we can which will tell us something about the next numbers. Some possibilities could be:

(i) Add on 7 and then add on 8...

28 *Mathematics: Friend or Foe?*

(ii) The next number will be even, and the next also even. (Why?)
(iii) The next number will be divisible by 7 and the next by 9. (Why?)
(iv) The next two numbers added together will make the square of an even number. (Why?)
(v) The sum of 21 and the next number will be the square of an odd number (Why?)
(vi) $1 \times 3 \rightarrow 3$
$3 \times 2 \rightarrow 6$
$6 \times 1\frac{2}{3} \rightarrow 10$
$10 \times 1\frac{1}{2} \rightarrow 15$
$15 \times 1\frac{2}{5} \rightarrow 21$
$21 \times 1\frac{1}{3} \rightarrow 28$
$28 \times 1\frac{2}{7} \rightarrow 36$

The series $3, 2, 1\frac{2}{3}, 1\frac{1}{2}, 1\frac{2}{5}, 1\frac{1}{3}$...seems to be approaching 1. Will it ever get there?

(vii) Within any group of three consecutive numbers, two of them are divisible by 3 and the third has digits which add up to 10. (Why?) And so on.

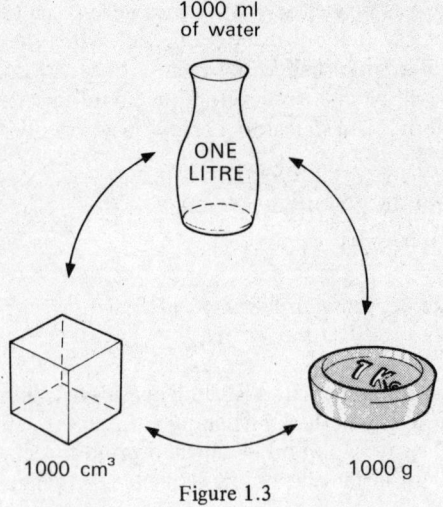

Figure 1.3

In another form of mathematics—measurement—it is pleasing to realise that there is an easily established link between measures of volume, capacity and mass (of water); see Figure 1.3. (It must be noted, of course, that absolute precision in these comparisons depends upon other factors.)

What Some People Say

Similarly it is intriguing to investigate the behaviour of a pendulum and to realise that for moderately small swings neither the size of the mass on the string nor the amplitude of the swing, but only the length of the string affects the time of the swing – and what is more, a string which measures a metre will take just a second to complete one swing.

Many similar 'discoveries' can provoke interest and inquiry. The story goes that Archimedes himself was so fascinated by the discovery that the volumes of a cone, sphere and cylinder (all dimensions based on the same sized circle) are in the ratio 1:2:3 that he requested a drawing incorporating these three shapes be drawn on his grave.

Volumes :   1   :   2   :   3

And have you discovered that it is not only so that the *square* on the hypotenuse is equal to the sum of the squares on the other two sides of a right-angled triangle, but that it also works with 'the semi-circle on the hypotenuse...'?

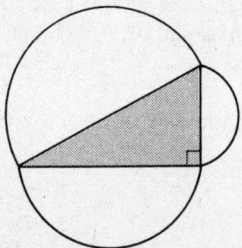

or 'the triangle on the hypotenuse...'? (the three triangles must be mathematically similar).

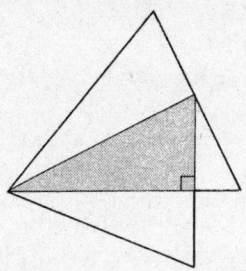

Might this also be true of *any* similar shapes on the three sides of a right-angled triangle?

Tessellations provide yet another absorbing (and, incidentally, useful) study. As well as considering the tessellating properties of quadrilaterals, triangles and the like, we may look instead at the techniques we can use to maintain the tessellation but destroy the original shape:

(i) By making corresponding patterns on opposite sides of the square (figure 1.4).

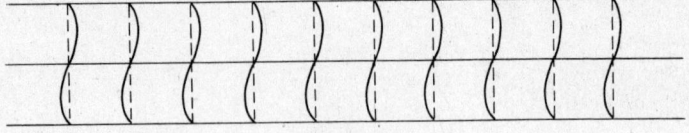

Figure 1.4

(ii) By making corresponding patterns on either side of the midpoint of each side (Figure 1.5).
(iii) By trying a mixture of both techniques (Figure 1.6).

As an extension of this, the work of M. C. Escher[1] provides an absorbing and profitable study. So too does a simple exploration of 'reptiles' (repeat tiles) in which a number of identical tessellating pieces

What Some People Say 31

Figure 1.5

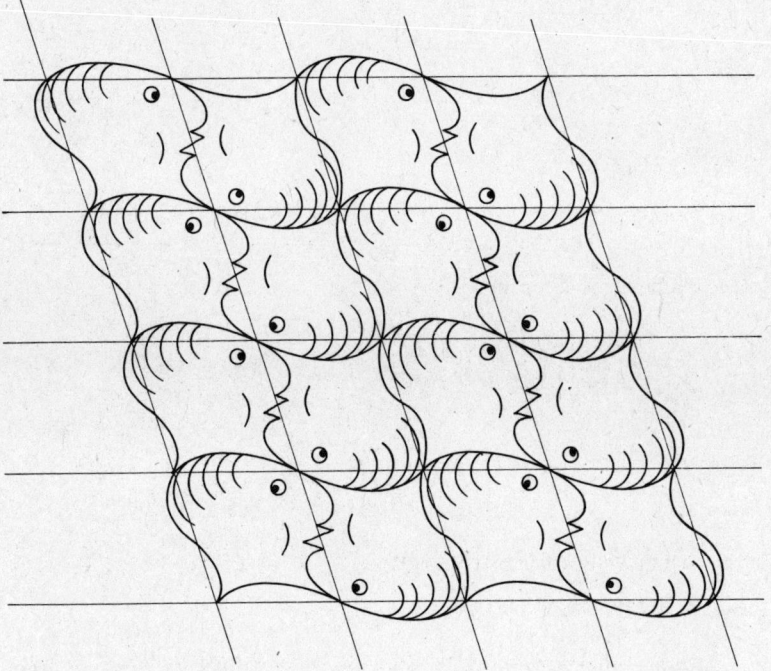

Figure 1.6

are required to form a shape which is (mathematically) similar to the original:

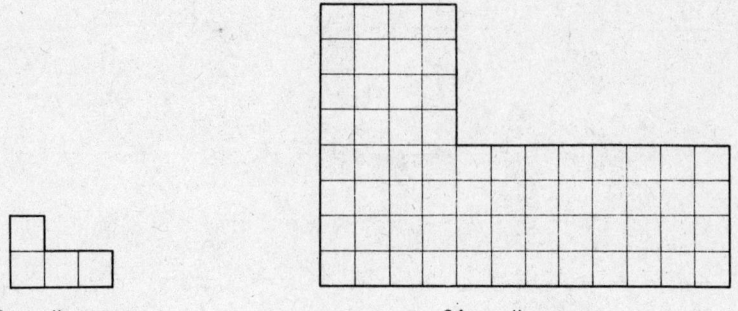

4 small squares

64 small squares

Can this L shape be constructed with less than 64 squares? It is easily seen that a square and a triangle are reptiles:

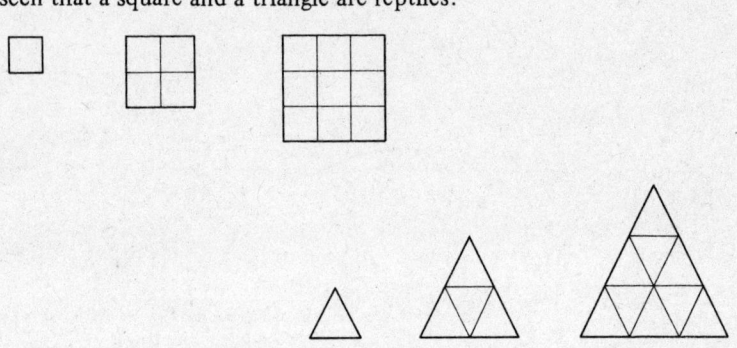

But is a hexagon a reptile?

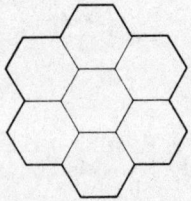

1.7   TECHNOLOGICAL ADVANCE

SOME PEOPLE SAY THAT:

> *'Mathematics helps us to keep abreast of the times, living as we do in an age of rapid technological advance.'*

What Some People Say   33

Perhaps this is an optimistic view of classroom mathematics but at least it is recognising the changing nature of the content of mathematics and the need for the teacher to be reviewing what is being included in the curriculum – and why.

One is reminded, when looking at the content of some school mathematics, of Harold Benjamin's satire[2] on curriculum entitled 'The Sabre-Tooth Curriculum'. In this satire, those people who attempted to change the original curriculum (consisting of fish-grabbing, woolly-horse-clubbing and sabre-tooth-tiger-scaring!) which had been designed to meet the former needs of the environment, to the present survival needs of net-making, antelope-snaring and bear-killing, met stern opposition from the wise old men who said: 'We don't teach fish-grabbing to grab fish, we teach it to develop a generalised agility which can never be developed by mere training.'

If one of the aims in teaching mathematics is to help to equip the child with the necessary skills to cope with 'everyday life' then the teacher must be willing to adapt his syllabus accordingly. This may require a mixture of foresight, courage and knowledge and an occasional readiness to make a leap in the dark. It is interesting to make a study of the history of mathematics[3] and to note that, as new demands arose, so man was able eventually to create the necessary mathematical tools to cope. For instance, when counting by tally methods became unwieldy, so the place value system of counting evolved. When large numbers had to be dealt with, so logarithms were invented, when rapid calculations were required, so calculating machines were invented, when power was harnessed to these machines, so computer programming techniques were produced. Occasionally in this development a Newton, a Napier or a Boole produced some mathematics which was ahead of his time and which spearheaded another step forward in the progress of mankind. In general it is true to say that schools have been cautious about introducing new topics related to some of these advances (more of this in Chapter 3). Many teachers, for instance, are still uncertain about the distinction between mass and weight, although the phenomenon of weightlessness is something which has been viewed on television by most young children. Number bases are still viewed with suspicion as being unnecessary or confusing, in spite of the fact that using them may not only clarify the structure of our denary system for us, but also give some idea of how a two-state system (e.g. on/off) can operate, as in a computer. Constructing or using a flowchart is still not generally considered to be sufficiently respectable to be included in the mathematics syllabus, even though looking at the structure of a process and breaking it down into its component parts is now considered to be a profitable way of understanding and retaining the process for future use. 'Playing' with apparatus such as Dienes Logic Blocks is still regarded by some to be wasteful of time, even though the child may be caused to think deeply and logically while attempting to solve a problem of classification or selection. The two examples in Figure 1.7, for instance, require a reasonable depth of thought on the part of an 8-year-old.

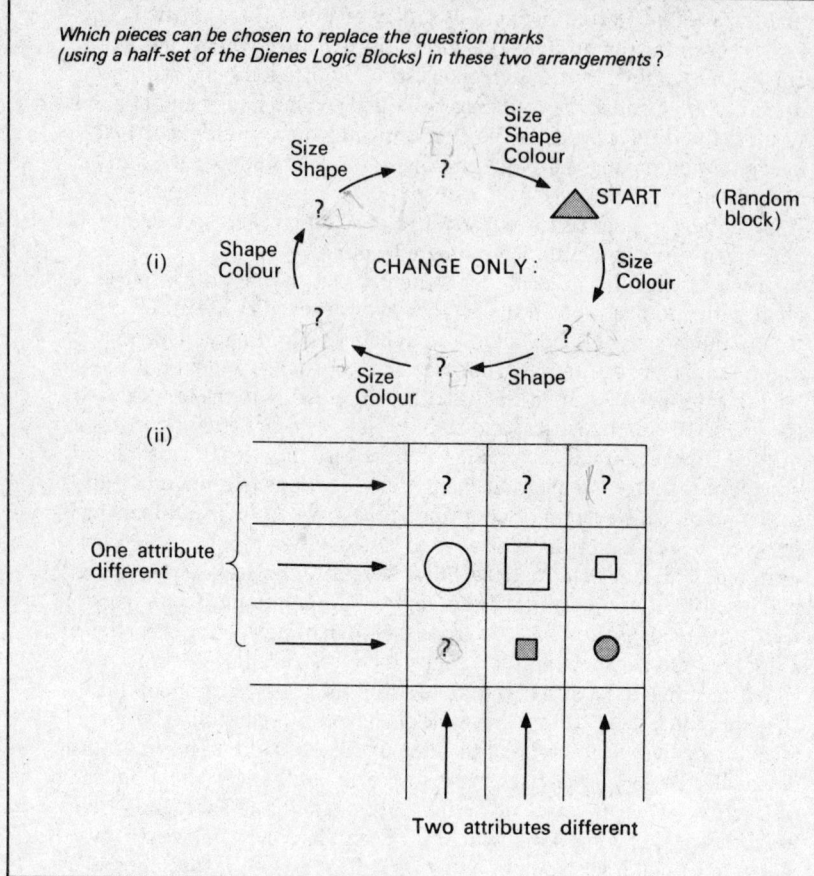

Figure 1.7

It is encouraging that recently many groups of teachers have been meeting to consider innovations in the curriculum, but meanwhile many children in our schools are being caused to do 'sums' which can only be shown to have merit as practice of techniques, rather than 'sums' which will help to produce thoughtful, resourceful learners, who can take a more informed interest in the achievement of their fellow men.

1.8 A MEANS OF COMMUNICATION

SOME PEOPLE SAY THAT:

'Mathematics supplies a basic core of knowledge which people expect each other to know and which provides a means of communication between them.'

Primary school mathematics is principally concerned with relationships within and between three main areas of study: number, measurement and shape. Within the first of these, number, we aim to equip a person with the vocabulary and a working knowledge appropriate to his needs of the numbers which exist and of what can be done with them. He is thus equipped with another form of language which he can use to communicate with other people. He understands what is meant by *21* January, or *21* London Road, or three sevens make *21,* or 'aged *21'*.... He is able to make any necessary calculation or to detect errors, he can appreciate the relative size of numbers in their context and to trust his own reactions when he registers surprise or horror or belief or satisfaction or acceptance of a numerical statement.

Similarly a working knowledge of measurement, 'the link between mathematics and the real world', is something which an adult tends to take for granted. Like number, measurement has an undisputed place in the mathematics curriculum. Having understood the concept of measurement as a form of comparison of one unit of measure with the 'object' being measured, we are able to read with more understanding newspaper reports of which the following is typical:[4]

2 February 1975

An agreement signed in Brussels yesterday by Mr Peart, Minister of Agriculture, to buy sugar at £260 a ton for 18 months from the cane-growing Commonwealth countries, will put up the price in the shops by around 1p for a 2lb bag. The present average retail price is 29p to 30p for 2lb.

There is still some argument between Mr Peart and the Commonwealth suppliers on exactly how much sugar Britain will get over the first six months of the agreement. Britain wants 450,000 tons but the producers say only 300,000 tons are available.

Edith Biggs stressed this need for understanding the concept of measurement when, to the somewhat apprehensive readers of 'Metrication in the School Curriculum'[5] she assured them 'centimetre or inch, litre or pint, kilogram or pound, square decimetre or square foot, what does it matter provided they [the children] understand the nature of measurement'. And, incidentally, she added encouragingly: 'There is an elegance and a simplicity about the metric system which need to be experienced before they can be fully appreciated.'

36 *Mathematics: Friend or Foe?*

Making a suitable use of space can be said to be partly innate and partly environmental. By studying shapes with the primary school child we are developing this spatial awareness through investigating what shapes exist, why they exist, how they are generated, where they arise naturally, what their properties are, what their uses and advantages are. This may lead not only to an economical and efficient use of space and its resources but also to an appreciation of the order and pattern within arrangements in the natural world.

To transmit and record the language of mathematics, symbolism has been developed. In particular this has provided us with an amazingly efficient numbering system (as mentioned on page 19). But to add to this we have an internationally known language consisting of such symbols as

$$+ \quad - \quad \times \quad \div \quad \sqrt{} \quad < \quad > \quad =$$

Similarly we are provided (thanks to mathematicians such as Descartes) with a system for recording the position of points in space:

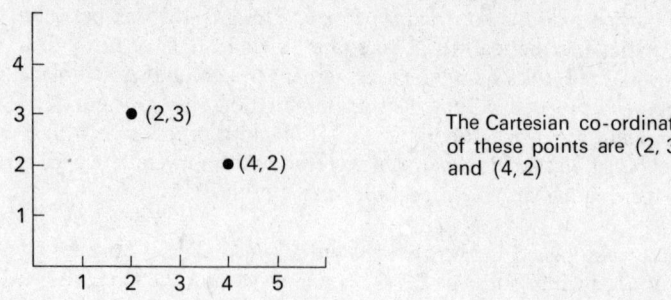

The Cartesian co-ordinates of these points are (2, 3) and (4, 2)

Using this simple device we can proceed to name a set of points in space:

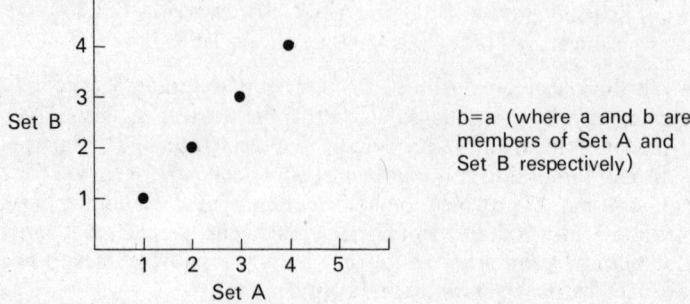

b=a (where a and b are members of Set A and Set B respectively)

We can further use a graphical device to communicate pictorially, for quick interpretation, a great deal of information, e.g. see Figures 1.8 and 1.9.

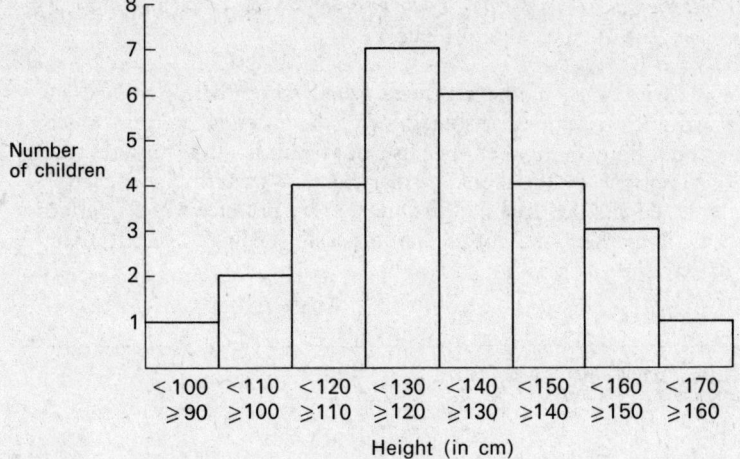

Figure 1.8

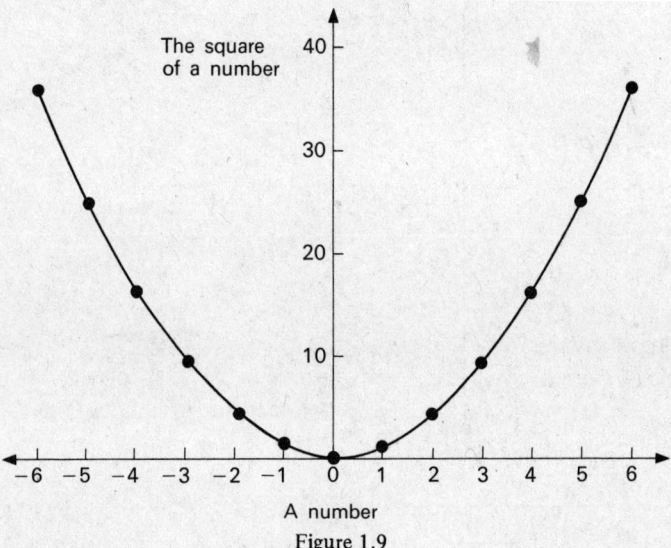

Figure 1.9

I wonder whether, as teachers, we tend to shroud comparatively simple symbolism in a mathematical mystique, and so undervalue some of the power of the language of mathematics?

1.9 A FORM OF BEAUTY

SOME PEOPLE SAY THAT:

> *'Mathematics is a form of beauty – and "a thing of beauty is a joy for ever"'* (with apologies to Keats!)

People who enjoy mathematics are usually those who have had an opportunity to appreciate something of the order and pattern which occur naturally in number or shapes or graphical work, for instance. In other sections of this book various number patterns have been mentioned and it is true that in almost every instance when a pattern occurs in mathematics nothing on this earth is going to stop it! (See Figures 1.10 and 1.12.)

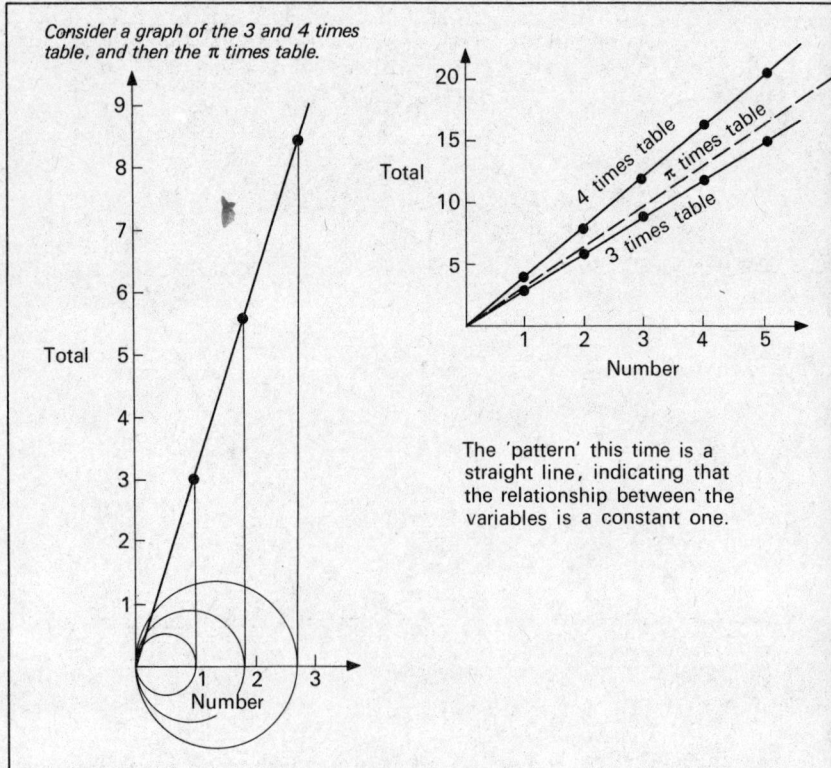

Figure 1.10

The form of beauty is not necessarily visual. It may be numerical. For instance, picture some cakes on a plate, from which *any* selection can be made. A number pattern emerges (Figure 1.11).

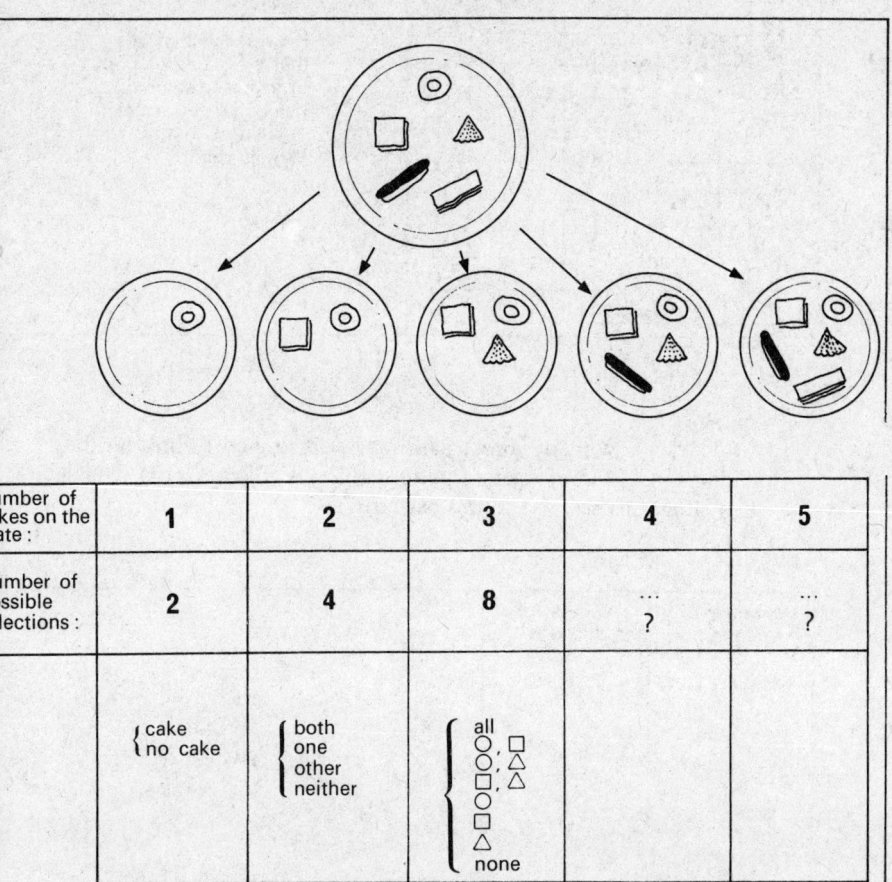

Figure 1.11

40 *Mathematics: Friend or Foe?*

Where are all the 12s in this tables square?

1	2	3	4	5	6	7	8	9	10	11	12
2					12						24
3			12				24				
4		12			24						
5											
6	12		24								
7											
8		24									
9											
10											
11											
12	24										

Where are the 24s?

How satisfactory it is that they are lined up in a graceful curve (including those not actually written down). Why is this?

Figure 1.12

Consider how many direct pathways are required to link up the buildings in a village. Again a pattern emerges (Figure 1.13). Which other situations give rise to this pattern?

Number of buildings	Number of pathways
1	0
2	1
3	3
4	6
5	10
6	?
7	?

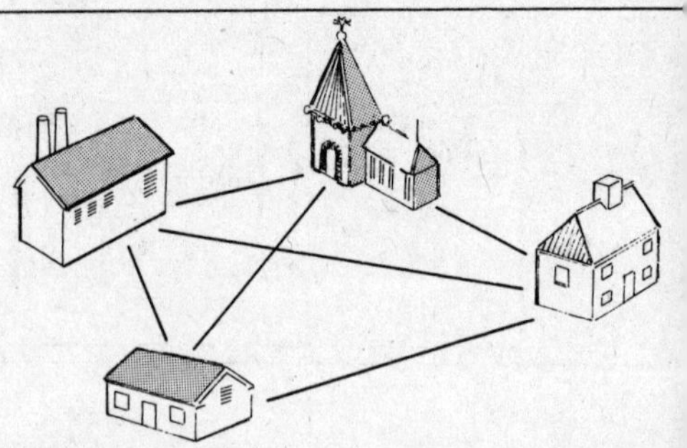

Figure 1.13

It is sometimes difficult (and unnecessary) to distinguish between the art and the mathematics displayed in the primary school classroom, in particular when different kinds of symmetry have been the subject of

What Some People Say 41

the work. Objects which have a form of symmetry have their own kind of beauty:

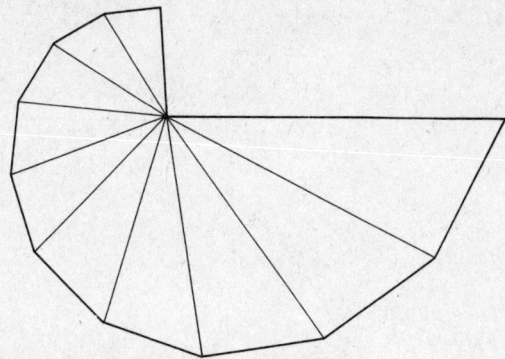

rotational symmetry   bilateral symmetry   both bilateral and rotational symmetry

Spirals provide an intriguing mathematical enquiry – as in the case of the equi-angular spiral such as the shape of the Nautilus shell (Figure 1.14).

Figure 1.14 *Diagrammatic view of the Nautilus shell.*

In the double spiral, which occurs frequently in nature, the number of spirals radiating clockwise and anti-clockwise can usually be counted (Figure 1.15). It is an intriguing fact that these two numbers are consecutive ones in the Fibonacci series:

$$1, 1, 2, 3, 5, 8, 13\ldots$$

When finding all the points in space which obey a given 'rule' we frequently create a thing of beauty. The parabola arises when the rule is 'keep equidistant from the fixed point and the fixed line' (Figure 1.16).

It is not difficult to find both hidden and apparent beauty in mathematics! Apart from the visual appeal of these examples quoted, the mathematician will find a hidden beauty in the simplicity and sophistication of the place value system, in the 'elegance and simplicity'

42 *Mathematics: Friend or Foe?*

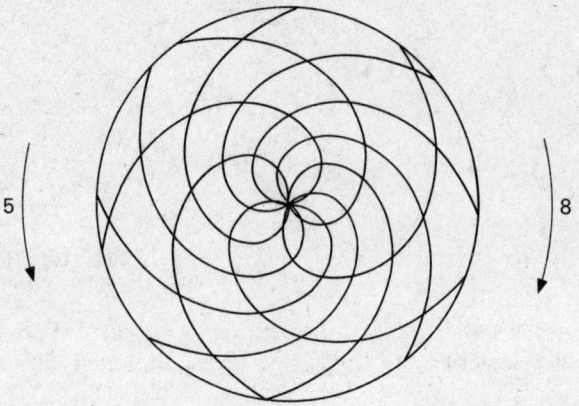

Figure 1.15

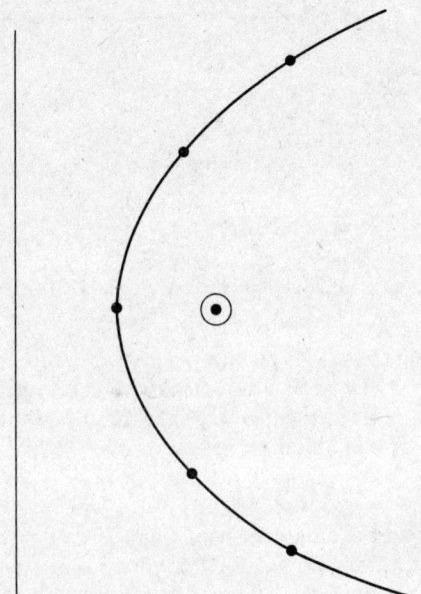

Figure 1.16

(to quote E. Biggs[6]) of the metric system, in the smooth efficiency of the logarithmic system, and in countless other forms.

1.10 ORDER OUT OF CHAOS

SOME PEOPLE SAY THAT:

'Mathematics provides a way of bringing order out of chaos.'

L. G. W. Sealey[7] re-echoes the statement that 'One of the most important things said about mathematics is that the patterns of mathematical thinking are much the same as the fundamental patterns of all thinking. If this is so, then it is obvious that mathematical learning assumes great importance.'

The problem for the teacher becomes that of helping the child to think mathematically. Could this start by bringing order out of chaos?

Sorting and classifying are activities which are familiar in the first school classroom, but are also of great importance in any problem-solving situation.

An 'everyday' problem may begin with an accumulation of seemingly unrelated and (possibly) redundant facts. The need to sort and classify the available data and thereby to find any necessary relationships to solve the problem is a first priority. If, by studying mathematics, we encourage an orderly way of bringing order out of the chaos, thus providing the beginnings of a strategy for problem solving, then mathematics has served a useful purpose.

Suppose we have:

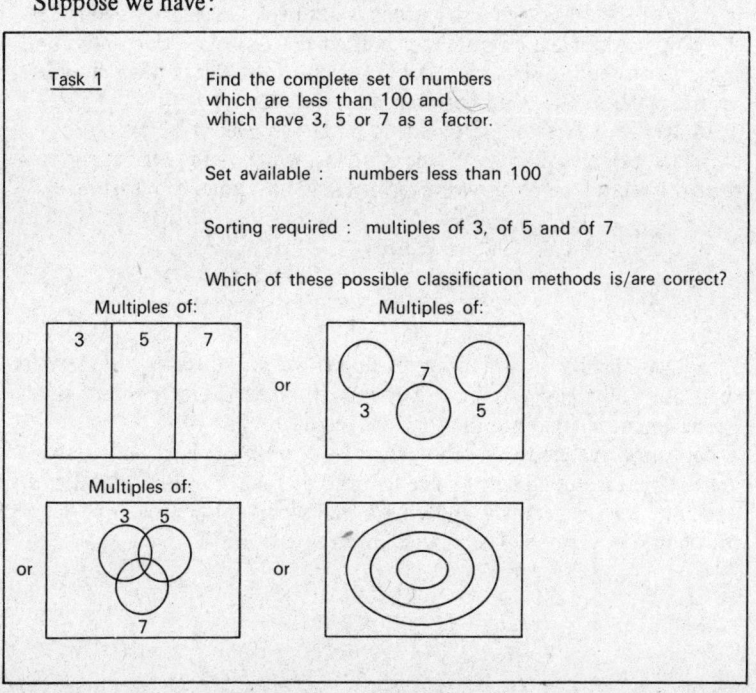

44 Mathematics: Friend or Foe?

Task 2   Choose suitable subsets of the (half-set) Dienes Logic Blocks to use the following classification methods:

Set available: Dienes Logic Blocks (half-set)

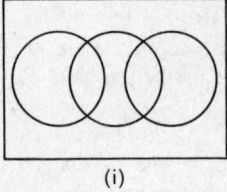 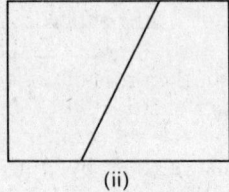 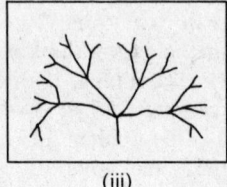

(i)   (ii)   (iii)

Sorting required for:

(i) three subsets, two of which are disjoint, the third intersecting with both

(ii) two complementary subsets

(iii) 4 categories × 3 categories × 2 categories

In each of these contrived situations it has been necessary to consider the properties of the original unclassified set, to isolate given required subsets and then to select a suitable form of classification which will maintain the relationship between these subsets.

Multiplication tables provide us with one means of classifying numbers. So, too, do measurement tables, which emphasize at the same time the relationships between various standard units of measure:

$$\text{e.g. } 1\,000 \text{ millilitres} = 1 \text{ litre}$$
$$10\,000 \text{ cm}^2 = 1 \text{ m}^2$$
$$1\,000 \text{ grammes} = 1 \text{ kilogramme}$$

If we can classify a number according to several different properties then our knowledge of that number may be sufficient to meet our needs whenever that number occurs (see page 17).

To use a systematic method to solve a problem is usually less vulnerable (although sometimes less rapid) than the use of trial and error. In this respect, by applying a mathematical approach to a problem there may be less chance of error:

Question: What is the probability of scoring 7 when two dice are thrown together?
Ad hoc method: To score 7 I could get:
(4,3) (1,6) (5,2) (3,4) (Have I got them all yet?)
What other scores might I get?
For 6 (1,5) (4,2)...
For 5 (2,3) (4,1)...
Have I exhausted all possibilities?
Systematic method: We will tabulate (classify) all possible results.

Score on second die

Score on first die

+	1	2	3	4	5	6
1	2	3	4	5	6	7
2	3	4	5	6	7	8
3	4	5	6	7	8	9
4	5	6	7	8	9	10
5	6	7	8	9	10	11
6	7	8	9	10	11	12

It is clear to see that the probability of scoring 7 is $\frac{6}{36}$ and, without doubt, we have omitted no other possibility.

1.11 LOGICAL THINKING

SOME PEOPLE SAY THAT:

> 'Causing a person to think mathematically is causing him to think logically.'

Perhaps this is merely a continuation of the previous section. Logic is becoming an accepted part of mathematics and the popular demand for Dienes Logic Blocks in the primary school classroom bears witness to the fact that many teachers believe it is important to encourage logical thinking.

The uninitiated person entering a classroom full of 9-year-olds who are 'playing' with Dienes Logic Blocks might be tempted to think that no mathematics was in progress. Taking a closer look we may find the children thinking deeply about:

(i) Arranging twenty-four blocks in an elaborate matrix pattern according to their own classification such that when a few pieces are removed other children can detect what has disappeared.

(ii) Lining up the full set of forty-eight blocks into a continuous circuit such that each block is two attributes different from its two neighbours.
(iii) Creating a symmetrical pattern and at the same time making sure that no piece is adjacent to a block which is less than two attributes different from itself.
(iv) Completing an 'Odd man out' exercise by classifying a half-set (twenty-four pieces) in the star formation illustrated in Figure 1.17.

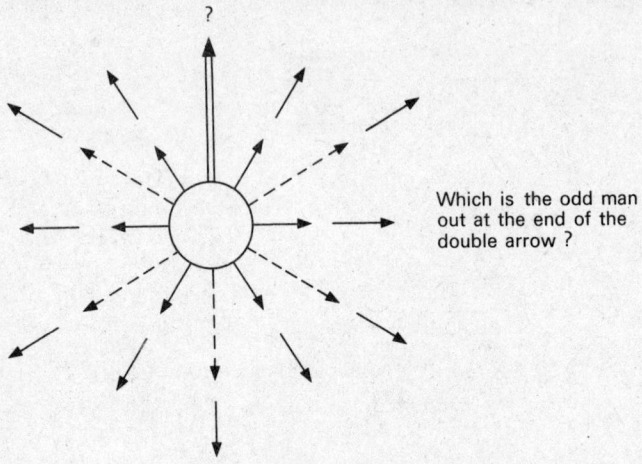

Figure 1.17

From a random piece in the middle, other blocks are placed such that at the end of a continuous arrow the piece is one attribute different from the block at the beginning of that arrow. At the end of a dotted arrow there must be a two-attribute difference.
(v) Playing Cat and Mouse (using only *two* attributes e.g. shape, colour; ignore size, thickness; Figure 1.18). Each cat remains still and is ready to pounce on any mouse which comes horizontally in line with himself and which has a similar attribute to himself, and which is *not* protected by another mouse. Each mouse moves one place at a time (forwards, diagonally or sideways) past the line of cats, taking care to avoid being exposed to a cat which is dangerous to him.

While the Dienes Logic Blocks provide an excellent source of situations in which one can practise logical thinking, perhaps we need to consider whether it is possible to *train* a young child to think logically. My belief is that this may be so if each of the following stages is passed. Perhaps it is best to consider these as rungs of a ladder (Figure 1.19).

What Some People Say 47

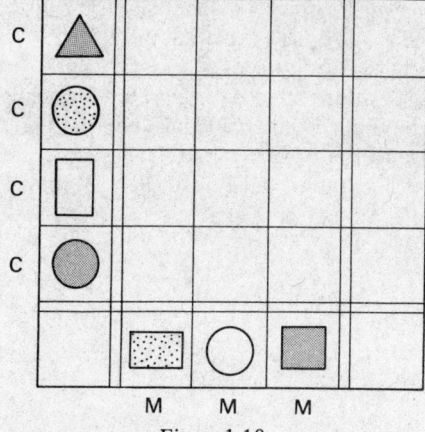

Figure 1.18

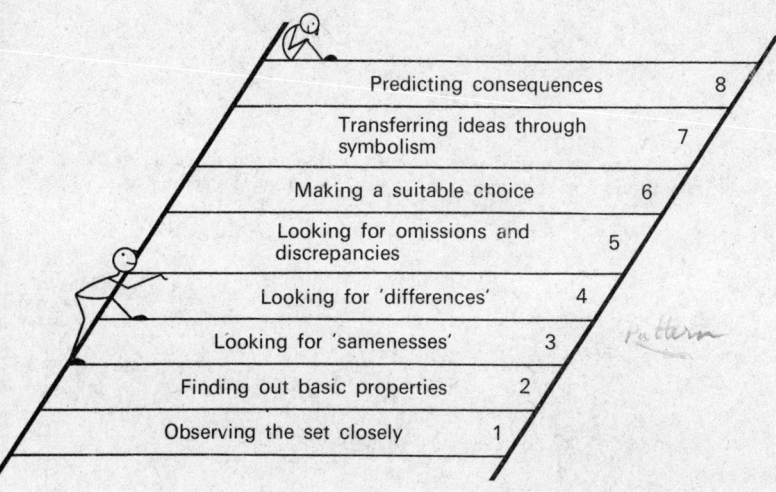

Figure 1.19

Rung 1 requires learning how to observe a situation closely - what objects are there, many or few, what colours, shapes, fabrics, special qualities (e.g. shiny, rough). The purpose of this is to cause the child to take an overall view of the extent and some possibilities of the available data. For this stage a random set which includes about ten unique objects with unambiguous descriptions is suitable to promote discussion, establish vocabulary and encourage close observation.

This leads to *Rung 2* on which the child may acquire knowledge of

48 *Mathematics: Friend or Foe?*

the basic properties of the objects individually. He will find out which ones will roll, or float, or squash, or bend, or feel hard....

Effective, logical thinking needs to be developed from basic knowledge of the situations about which we are reasoning. With young children it might be that sorting, matching and classifying activities will increase knowledge of the set being considered. See, for example, the matching activities illustrated in Figures 1.20, 1.21 and 1.22

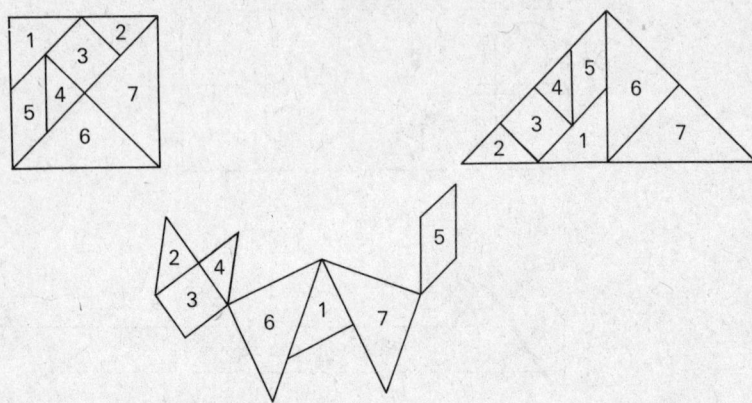

Figure 1.20 *The seven pieces of the tangram may be matched on to outlined skeleton diagrams (not yet on to silhouettes as for normal tangram puzzle).*[8]

Figure 1.21 *Pattern cards may be matched on to appropriate cells.*

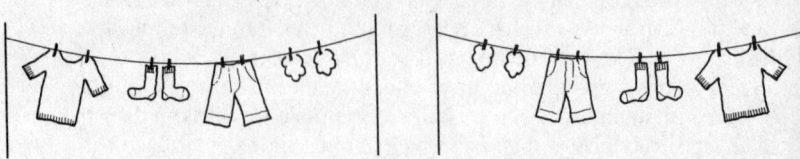

Figure 1.22 *A reversal of positions may be required.*

What Some People Say

Rung 3 could be called 'looking for samenesses', the purpose being that some appropriate criterion is found which will group or, possibly, partition the set. This can be encouraged perhaps by 'conversations' between objects in which a remark may be exchanged which is true no matter which object says it:

e.g.

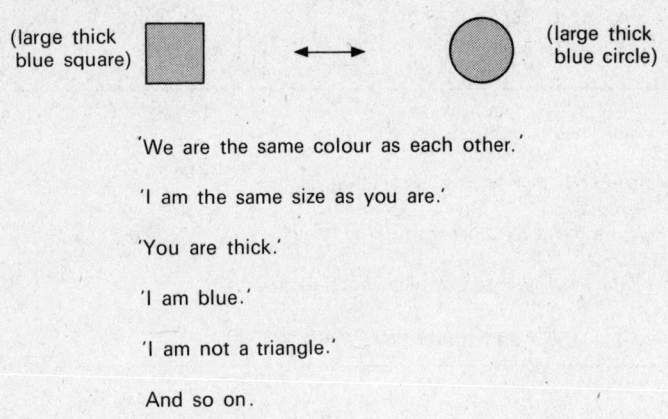

'We are the same colour as each other.'

'I am the same size as you are.'

'You are thick.'

'I am blue.'

'I am not a triangle.'

And so on.

To encourage observation of samenesses the completion of a matrix pattern is appropriate, using the thick Dienes Blocks:

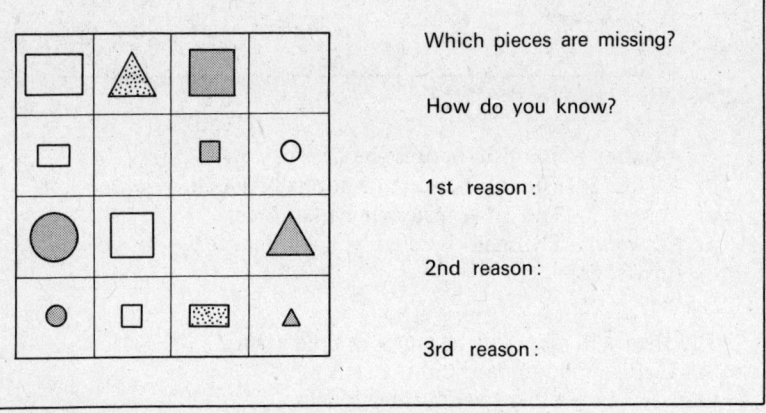

50 *Mathematics: Friend or Foe?*

Or a set of pictures may be scrutinised for samenesses:

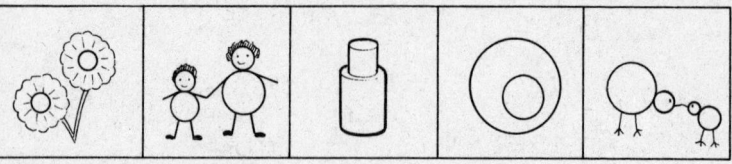

'Each picture has two things.'

'Each picture has at least two circles in it.'

'Each picture has a red thing and a blue thing.'

'Each picture has two parts which touch each other.'

'Each picture is on a card which is the same size.'

It is to be expected that *Rung 4* of the ladder is 'looking for differences'.
Which is the odd man out? Why?

Answer 1: The fish - because he lives in water.
or Answer 2: The cat - because he normally lives in a house.
or Answer 3: The cat - because he has a fur coat.
or Answer 4: The snail - because he has antennae.
or Answer 5: ...

The Dienes Blocks (half-set) may be used again:
Start with a random block in the centre.
Classify all the other twenty-three blocks.

What Some People Say 51

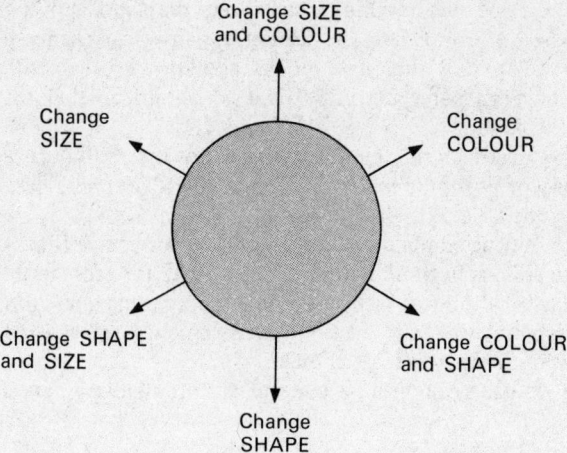

Or they may be arranged as the cargo on a train:

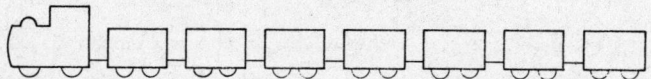

Arrange the small yellow blocks so that each part of the train carries one piece which is different in one/two way(s) from its neighbours.

Rung 5 is called 'looking for omissions and discrepancies' and here the child is being required to look critically at the situation to find out what can be done to put things right:

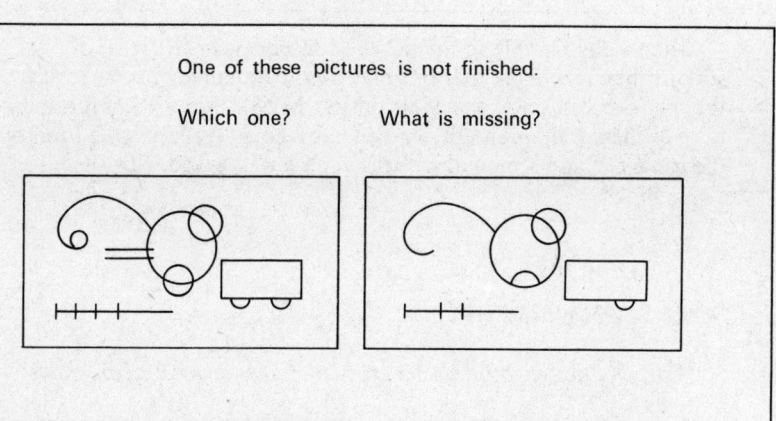

Scatter the Dienes Blocks (full set) on the table. Hide five pieces. By process of elimination decide which pieces are missing.

52 Mathematics: Friend or Foe?

With this same idea in mind two children can play Happy Families, having shared out, in a random fashion, a half-set of Dienes Blocks. Being able to see only their own blocks, and knowing the contents of the set, they can name correctly (and so obtain) one of their opponent's pieces.

Rung 6, 'making a suitable choice', requires the child to survey the whole array of possibilities, eliminate the redundant ones, classify the remaining one, select the right one(s).

Perhaps, having climbed successfully the previous five rungs of the ladder, the child is beginning to devise a strategy for problem solving? Using a classification tree (see page 44) or completing a circuit (see page 46) or completing the 'odd man out' arrangement (see page 46) will test the ability to 'make a suitable choice'.

Ability to use symbolism, a code or an intermediary constitutes *Rung 7.*

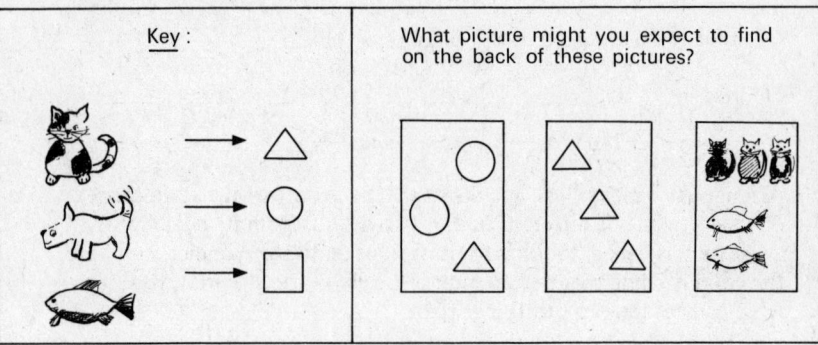

When a child is able to predict the consequences (*Rung 8*) of his actions then is it likely that this may be the outcome of some logical thought processes. For the adult this might be epitomised in the ability to play chess well. A child's game of chess could well take the form of the game Cat and Mouse described on page 47, played with Dienes Logic Blocks.

1.12   ACCURACY

SOME PEOPLE SAY THAT:

> *'Mathematics encourages accuracy and should rank as an exact science.'*

This claim for mathematics is an important one but it should be recognised that while exact accuracy is sometimes required for many situations, it is neither possible nor necessary (nor even desirable) that a precise answer is always forthcoming. Where absolute accuracy can be expected in the junior school classroom is usually in computation and in counting. However, even with these activities, an estimated or approximate answer is sometimes more useful and is certainly quicker to obtain:

How long is this room? This may produce many answers:

> It is 40 and a bit tiles long.
> It is about 10 metres long.
> It is 10 metres and 3 centimetres long.
> It is 15 strides long.
> It is twice as long as the rope in the cupboard.
> It is 10.32 metres long.

Which answer is right? In a given context, each one can be said to be the 'best' answer.

By encouraging a child to estimate an approximate answer the teacher is not only giving him a means of cross-checking his own calculated answer but also testing whether the child has acquired the concept involved. If a child who is multiplying 98 by 7 is able to say that the answer is somewhere around 700 then the teacher can rest assured that the child is on the right lines and is likely to proceed through the sum successfully.

It is said with some truth, that of all forms of measurement (length, mass, capacity, time, angular, temperature, area, volume...) the only one for which we can usually guarantee accuracy is counting—and that is the case only when the number being counted is finite and limited! We are dependent, for instance, upon our own errors, the manufacturer's errors in the making of the measuring instrument, the suitability of the instruments available for the exercise, the conditions at the time of the measurement. So some interesting and profitable discussion might arise about the accuracy of a certain child's answer that, because the length of the corridor is 20 metres 59 centimetres, his pet tortoise (assuming that it will travel obediently in a straight line without diversion) which walks at an average rate of 1 metre every 12½ seconds, will take 20.59 × 12.5 seconds to go the whole way!

So there is a place for accuracy and a place for approximation and estimation in our teaching of mathematics and the important thing is that the child should be encouraged to assess the appropriate degree of accuracy for himself, according to the nature of the task and the variables involved. Climbing the rungs of the logic ladder may help this process (page 47).

So possibly we should guard against creating only mathematical situations for a child which always (or almost always) result in a correct, unique answer. The experience of realising that there may be many suitable, but varying, answers, or even no answer at all, is both salutory and true-to-life!

When considering accuracy one marvels at the degree of precision achieved by mathematicians of the past. I. E. S. Edwards in *The Pyramids of Egypt*[9] tells us that 'No monument in Egypt has been surveyed and measured so often and with so much care as the Great Pyramid'. From a survey in 1925 it was found that the four sides of the base measured 755.43 feet, 756.08 feet, 755.68 feet and 755.77 feet - a difference of only 7.9 inches between the longest and the shortest. And what is more, the Great Pyramid faces due north, south, east and west with a maximum error of less than one tenth of a degree. Nevertheless, while one admires the accuracy indicated by these numbers, it is possibly the comment that 'the Houses of Parliament and St Paul's Cathedral could be grouped inside the area of its base and still leave considerable space unoccupied' which captures the imagination and is a suitably meaningful measurement for some purposes.

So perhaps the group of children in one school who found they were able to make a life-size model of a dinosaur in their hall, or the group in another school who discovered that if a dinosaur walked into their playground he would have to leave part of his tail sticking out through the gate, were doing some useful mathematics - admittedly not with the precision required of the designer of a space rocket!

1.13 INGENUITY AND RESOURCEFULNESS

SOME PEOPLE SAY THAT:

> *'Mathematics is a subject through which one learns to be resourceful, imaginative and ingenious.'*

People who have had a fair amount of mathematical experience acquire an inner resourcefulness and ingenuity because they are building up a store of 'mathematical models' or strategies or solutions which will stand them in good stead in the event of further problems arising requiring a solution. I have mentioned before (page 21) the rag-bag strategy of solving problems by finding the next best thing if the correct piece is not available. Polya, in the Preface to his book 'How To Solve It.,[10] encourages a student to seek back in his memory for a related problem in order to get a start with his new problem. We could quote again the example of the child who needs to cut out a square with some degree of accuracy (page 21). Similarly the older student faced with an involved algebraic fraction, for instance, does well to prevent himself from making errors by deciding 'What would I do if this were a simple numerical fraction instead?'

The art of using mathematics successfully includes developing an ability to generalise. In its simplest form this could mean:

Since 2 cats and 3 cats make 5 cats altogether,
    2 books and 3 books make 5 books altogether,
    2 days and 3 days make 5 days altogether,

then perhaps we are 'safe' to assume that

$2 + 3 = 5$ usually

The inner resourcefulness which comes from being able to move from the particular to the general, however, is something which can be neither hurried nor forced. It is my belief that much of the unpopularity of mathematics with older children (particularly girls) is that they have been encouraged too soon to accept and to use methods which are too sophisticated and too generalised. Just as we would not expect a young child to paint with the skill of a mature artist, so we should not expect a child to learn through using short-cut mathematical processes which have been invented and perfected by someone who is better able to understand them and to see the reason for them. For instance the transition from:

$$\tfrac{3}{5} + \tfrac{3}{4} \rightarrow \tfrac{12+15}{20}$$

can be learned as a rule of thumb. It is easy to cash in on someone else's methods, but for real understanding the child must grasp for himself the concept of equivalence of fractions. It may be that some children do learn best by using a borrowed technique, finding out that it works and then afterwards finding out the reason why, but the important thing is that time and opportunity should be given for the understanding to come. Unless this is so the child will not gain confidence to venture from there into the unknown where resourcefulness and ingenuity need to be built upon basic understanding. W. Servais,[11] surveying mathematics in European schools, reports: 'The goal is motivating, constructing, understanding, developing an ability to arrive at correct decisions and to model simple situations.'

1.14   AN EXPLANATION OF NATURAL PHENOMENA

SOME PEOPLE SAY THAT:

> 'Mathematics is able to capture, record and sometimes explain what exists in nature.'

56 Mathematics: Friend or Foe?

There is certainly endless support for this claim for mathematics. The triangle, circle, ellipse, parabola, hexagon, cone, cylinder, cube, octahedron, dodecahedron, icosahedron, spiral, double spiral...all of these are natural phenomena. It is true that man copies and uses them, but nature gave him the pattern – and the most efficient uses for them as well!

Take the ellipse, for instance: a planet travels in an elliptical orbit and Kepler proved that the line joining the planet to the sun sweeps out equal areas in equal time (Figure 1.23). We can copy the ellipse by

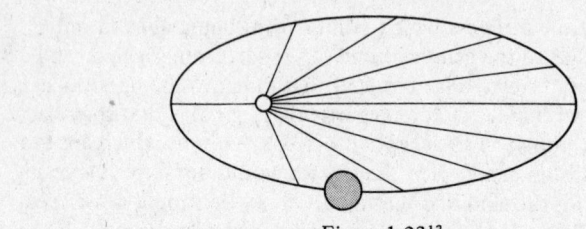

Figure 1.23[12]

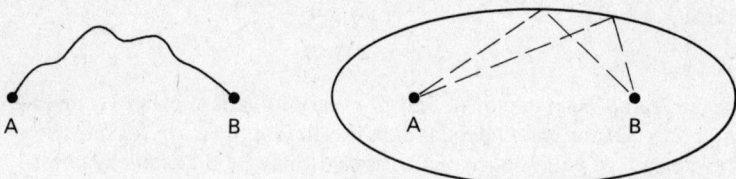

Figure 1.24

keeping constant the total distance from a moving point to two fixed points (Figure 1.24).

Using the system of Cartesian co-ordinates we can capture this curve in an algebraic equation, which expresses the general relation between the co-ordinates of the points on the curve.

Euler, the German mathematician, clarified a link between natural networks and mathematics. He interested himself in topological maps illustrating networks and he investigated the traversibility of these, having been motivated to do so, it is said, by the famous problem of the Seven Bridges of Königsberg. Euler realised that a network can only be traversed if the number of odd vertices (junctions in which an odd number of roads meet) is either zero or two. So can these be traversed?

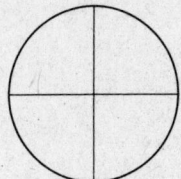 

Euler, using a topological extension of this idea, mentioned the relationship between the numbers of edges, faces and vertices of the platonic solids.

Platonic solid	Faces	Edges	Vertices
tetrahedron	4	6	4
cube	6	12	8
octahedron	8	12	6
dodecahedron	12	30	20
icosahedron	20	30	12

giving us the somewhat remarkable equation:

$$F + V = E + 2$$

1.15 A STUDY OF SETS

MOST PEOPLE ADMIT THAT:

'Mathematics is a study of sets.'

I have put this claim last as it seems to me to be the most comprehensive of them all. On page 26 we considered the claim that mathematics provides us with a way of looking at the physical world. Within that physical world the mathematician is particularly interested in sets of numbers, sets of measures, sets of shapes, sets of points in space, sets of articles and sets of people.

At the risk of compartmentalising these ideas too neatly and artificially, the progressive stages of development could be:

(i) seek properties within the one set, i.e. establish subsets;
(ii) seek relationships between the subsets of the one set;
(iii) seek relationships between two, or more, sets.

Using this explanation, primary school mathematics would classify itself neatly (too much so?) according to Figure 1.25 in which the

58 *Mathematics: Friend or Foe?*

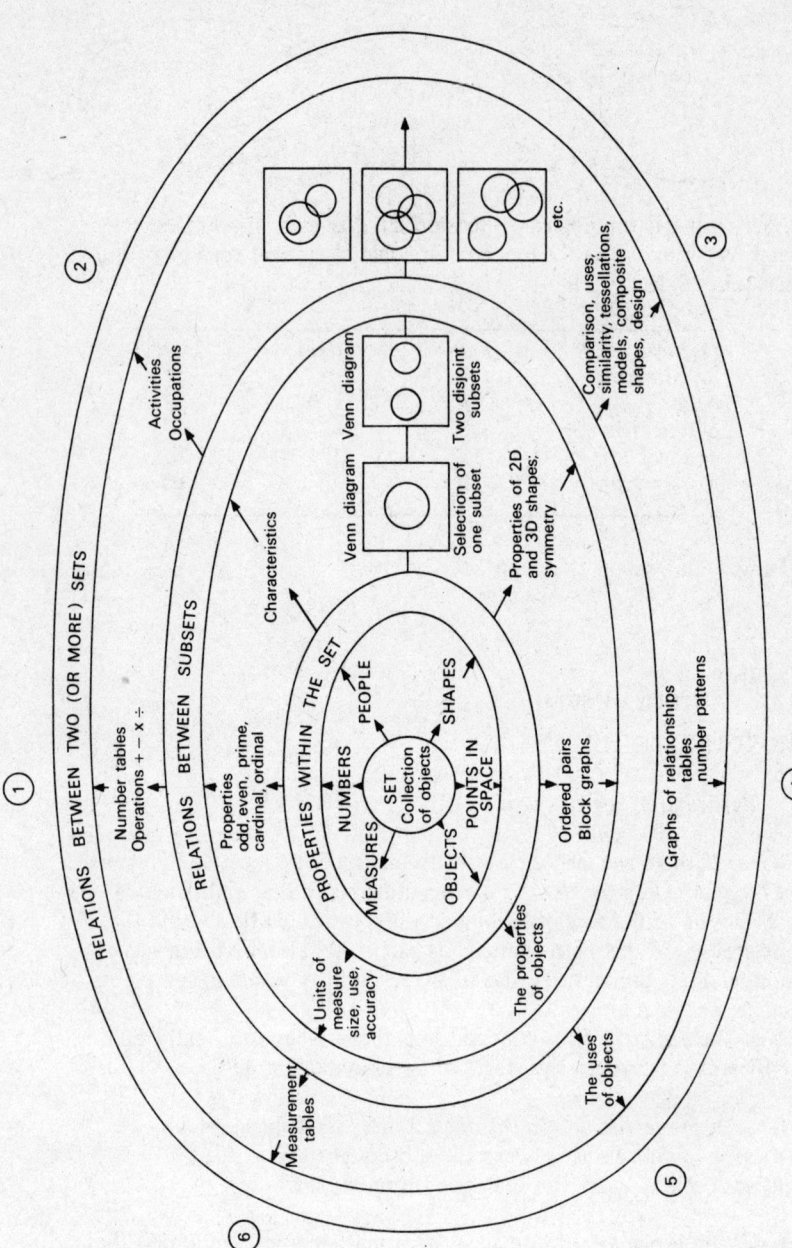

Figure 1.25

spokes indicate study of the main sets during stages (i) and (ii). The interrelationships of stage (iii) between any set and any other provide the major content of the mathematics we teach:

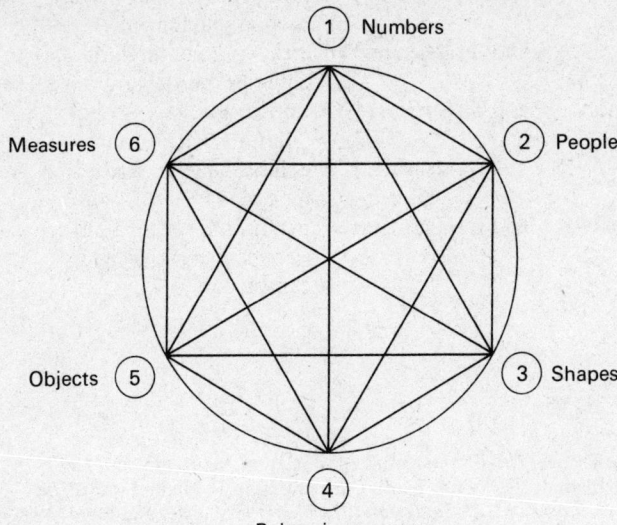

For example, mathematics arising from:

1 (Numbers)	and 2 (People)	→	statistical data
1	and 3 (Shapes)	→	triangular numbers, number patterns, clock patterns
1	and 4 (Points)	→	graphical representation, equalities, inequalities
1	and 5 (Objects)	→	Counting, quantities, dice projects
1	and 6 (Measures)	→	measurement tables, ratio, scale
2 (People)	and 3 (Shapes)	→	design, efficiency, economy, usefulness
2	and 4 (Points)	→	frequency graphs, normal distribution, data recording
2	and 5 (Objects)	→	trade, money, transactions, surveys
2	and 6 (Measures)	→	heights, weights, records, shopping

3 (Shapes)	and 4 (Points)	→	straight and curved line graphs, ellipse, parabola, circle
3	and 5 (Objects)	→	3D shapes and properties, natural phenomena
3	and 6 (Measures)	→	area, volume, similarity, circle properties...
4 (Points)	and 5 (Objects)	→	transformations, relations matrix chart
4	and 6 (Measures)	→	graphs, heights, ready reckoner
5 (Objects)	and 6 (Measures)	→	pendulum, speed, area, volume, length, weight, capacity

REFERENCES

1 M. C. Escher, *The Graphic Work of M. C. Escher* (Macdonald, 1967).
2 H. Benjamin, 'The Sabre-Tooth Curriculum' In R. Hooper (ed.), *The Curriculum: Context, Design and Development* (Oliver and Boyd, 1971).
3 See, for example: H. A. Shaw and K. Fuge, *The Story of Mathematics* (Arnold, 1963); M. Holt and A. J. McIntosh, *The Scope of Mathematics* (OUP, 1968).
4 Extract from *The Daily Telegraph*.
5 E. Biggs, 'Metrication in the School Curriculum' in *Trends in Education*, No. 26 (April 1972).
6 E. Biggs, op. cit.
7 L. G. W. Sealey, *The Creative Use of Mathematics in the Junior School* (Blackwell, 1966).
8 See M. Gardner, *More Mathematical Puzzles and Diversions* (Pelican, 1966).
9 I. E. S. Edwards, *The Pyramids of Egypt* (Pelican, 1961).
10 G. Polya, *How To Solve It* (Princeton University Press, 1957).
11 W. Servais, article in *Mathematical Education in Science and Technology* (February 1975).
12 D. Bergamini, *Mathematics* (Life Science Library, 1970).

Chapter 2

The Syllabus

2.1 AN *AD HOC* OR STRUCTURED APPROACH?

Every teacher of mathematics faces a dilemma, a challenge, a problem - call it what you will - with regard to the curriculum. The word curriculum in this context is used in its widest sense to include overt and hidden aspects, so it embraces not only the content and organisation but also the style of working and attitude of the people concerned.

This part of the book is intended to put on display a representative selection of ideas about content for inclusion in a scheme of work, to explore similarities and differences between these ideas and to attempt to find a useful pathway running through them.

Very few teachers are prepared to commit themselves to a definite sequence of topics to be included in a mathematics syllabus. I do not think that this hesitancy is wrong because there is no sequence which is right for all children, all topics, all teachers and all circumstances. So in the end it is only the teacher who is able to feel his way through a certain section of the work. However, while supporting this opinion, I have no hesitation in saying that:

(a) Many of these people have committed themselves to paper and in doing so have implied certain priorities and sequences of work.
(b) When a series of teachers are responsible for a child's learning in mathematics it is essential that there should be some plan which ensures that the child has covered the core of essentials in a reasonably coherent and understandable way. This plan may be pre-determined by:
   (i)   a nationally used syllabus;
   (ii) a syllabus devised within one authority area;
   (iii) a syllabus devised by the headteacher or leader of mathematics in the school;
   (iv) a textbook;
   (v)   an evaluation scheme consisting of a record book, or progress guide or examination.

Teachers' opinions on how the scheme of work should be devised vary as much as the above list. Their comments include:

'The syllabus is not a one-man job – the whole staff must be involved in discussion about it.'

'The primary schools *want* discussion with (name of a secondary school) about the work we do but there is no consultation except that forced by us.'

'Our main contact is through feedback from children who go on there!'

'In a county we should all have a system – we must have continuity.'

'We have no syllabus as such. Everyone picks out what they like. We have fortnightly mathematics meetings for all the staff and we share ideas there. Also each teacher keeps a record book which is handed from teacher to teacher indicating the sections of the textbook (usually Fletcher) covered by individual children.'

'The teachers of the two infant classes have their own mini-syllabuses. The rest of us are trying out our own individual schemes and are intending soon to discuss a plan for the school. The headmaster doesn't like much assessment or grading. Some of the less able children have blossomed but we hope the able have not lost out through no competition.'

'We base our work on Fletcher, supplemented by the ITV programmes "Figure It Out" for the lower juniors and the BBC "Maths Workshop" programmes for the upper juniors. Occasionally we dip into other textbooks for workcard ideas.'

'Basically, it is based on Fletcher, especially with the infants. Level II is not so good though. He does go on a bit! The sequence suggested doesn't tie in with other schemes. He is the odd man out. Fletcher puts flesh on the mathematical bones, but the bones are in an unusual order!'

And so on....

One is left with a feeling of unease after assessing the pros and cons of an *ad hoc* approach:

PROS	CONS
Scope for:	*Danger of:*
Immediate relevance for the moment	Lack of coherence with programme of work
Interest	
Enjoyment	Superficial treatment of too many topics

Suitability

Motivation

Topics tackled too soon or too late
Incorrect learning order
Timing of progression difficult

Possible difficulties:
Assessment of progress
Children feeling insecure
Strain on the teacher of deciding 'where next'
Anxiety of parents

In the hands of a series of good teachers of mathematics the cons would disappear but, regrettably, this is a utopian situation. Many good, dedicated teachers would be the first to admit that they haven't the confidence or mathematical ability to cope in an *ad hoc* situation and therefore they seek the back-up of a more structured syllabus.

It is with this firm belief in mind that I am presuming to investigate a number of existing schemes in this part of the book. The schemes included are far from exhaustive but the brief analysis may help the teacher who is bewildered by a welter of seemingly conflicting ideas and schemes to match his own ideas with those of other people.

2.2 SCHEMES OF WORK TO BE ANALYSED

The schemes contributing to part or the whole of primary mathematics which are included are:

1. Mathematics Department of Manchester College of Education, *Guidelines in School Mathematics* (Rupert Hart-Davis, 1969).
2. H. Fletcher, *Mathematics for Schools, Level I* and *Level II* (Addison Wesley, 1971).
3. A scheme devised by a county authority.
4. ATCDE Mathematics Section, *Children Using Mathematics* (OUP, 1973).
5. Mathematical Association, *Primary Mathematics, A Further Report* (Bell, 1970).
6. L. G. W. Sealey, *Beginning Mathematics, Books 1-4* (Blackwell, 1968).
7. The Schools Council, *Mathematics in Primary Schools*, Curriculum Bulletin No. 1 (HMSO, 1965).
8. C. Stern, *Children Discover Arithmetic* (Harrap, 1953).

64 *Mathematics: Friend or Foe?*

9 Schemes devised with my colleagues and with local groups of teachers.

2.3 PLAN OF THE ANALYSIS

To make any useful sense out of this analysis I have run the risk of jeopardising the coherence of a totally comprehensive scheme by breaking it down into consideration of the following overlapping sections:

1 Number:
   Pre-number → Awareness of numbers to 100
2 Number:
   Operations: Addition
                Subtraction
                Multiplication
                Division
3 Number:
   Fractions, Decimal Fractions, Percentages
4 Measurement:

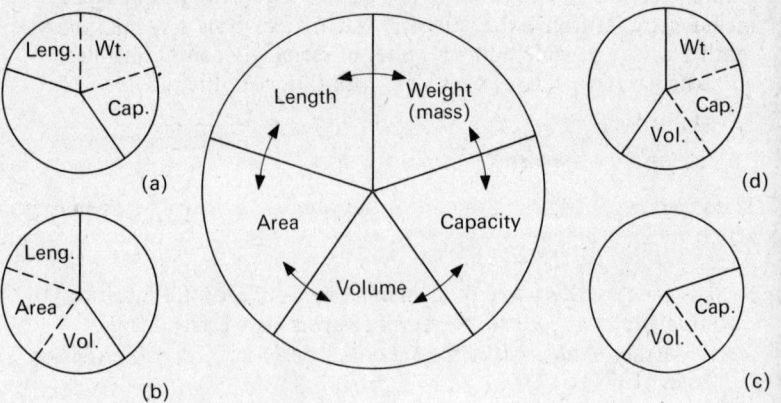

The related concepts of:
(a) Length, Weight (Mass), Capacity
(b) Length, Area, Volume
(c) Capacity, Volume
(d) Weight (Mass), Capacity, Volume
and also:
(e) Angles
(f) Time
(g) Money

5 Shape
6 Graphical Representation

Each of these sections is considered in turn. An attempt has been made to draw out what appear to be common opinions across some of the schemes and to discuss some diversions introduced by certain of them. It must be emphasised that this synopsis is based on my interpretation of the ways in which various schemes are presented and I apologise if they have been misrepresented in any way.

2.4 THE PRE-NUMBER STAGE TO AWARENESS OF NUMBERS TO 100

Sorting and matching are the first two activities introduced into all schemes as a basis for number work. While in most instances they provide a direct lead to *one-to-one correspondence,* in some schemes the concept of *comparison* in the size of groups is considered first. Sealey in particular compares group sizes and introduces the concept of conservation of group size before he comes to one-to-one correspondence. The *cardinal numbers, number names, numerals* and the *ordering of numbers* form the next major section of work in all schemes but the order in which these are introduced varies considerably. While some schemes deal immediately with numbers 1 to 10, others take numbers only up to 5 first. With regard to cardinal numbers and zero it is interesting to compare the differing approaches of Fletcher and Sealey. Fletcher deals with only the first five numbers, 2, 4, 3, 5, 1, and then proceeds to get them in order, before tackling 0, 6, 7, 8, 9, 10. No special remark is made when introducing 0 or 10. Sealey on the other hand introduces the first nine numbers in the order 2, 5, 3, 6, 1, 8, 4, 7, 9 and then proceeds to get them in order. Zero does not appear until grouping occurs in Book 3, and then the symbol for zero is made with no comment (Book 3, page 9).

The concept of *conservation of number* follows closely when the cardinal numbers are known and, in general, precedes consideration of the *ordinal numbers.* At this early stage some *symbolism* begins to appear in a few schemes, in particular symbols denoting comparison:

$$< \text{ and } >$$

The fact that each scheme has its own method of introducing *basic operations using the numbers from 1 to 10* demonstrates how widely people's ideas on this differ. While a county scheme favours counting on and back within the numbers 1 to 10, in order to establish addition bonds to 10 and number bonds to 6, Fletcher concentrates on addition by counting on up to 10. In this way he establishes the *addition* facts for 5 and 6, then 7 and 8, and then 9 and 10. The ATCDE report deals with number bonds up to 9 and then addition facts about 10, and introduces the symbols for addition and subtraction at this stage.

Grouping in 2s, 3s and 4s usually occurs early and is the first hint of forthcoming operations of multiplication and division of numbers. However, Fletcher does not do this until addition facts to 20 and counting on in 2s, 3s and 4s have been covered.

At this point most schemes concentrate on *numbers greater than 10 but less than 20,* as a forerunner to introducing ideas on *place value.* To introduce the notation for *numbers from 10 onwards* and the idea of place value, Sealey uses the grouping principle in 2s, 3s and 4s and higher numbers (e.g. Book 3, page 9).

Look at this group [⊚ ⊚ ⊚ ⊚ ⊚ ⊚ ⊚ ⊚]

We write its number: 8

Now we put the washers in twos.

[⊚ ⊚] [⊚ ⊚] [⊚ ⊚] [⊚ ⊚]

There are 4 twos and no ones left.

We can write this in a new way:

	twos	ones
8 =	4	0

Incidentally there are three absolutely new mathematical ideas for the child to meet in this one extract:

(a) the zero;
(b) the symbol for 'equals';
(c) the place value notation.

Fletcher adopts the 'bundles of 10' idea to introduce quantities greater than 10, the children recording ☐ tens ☐ units. In this way he takes in all the numbers from 10 to 100 at the same time (*Level I,* Book 4, pages 42–7).

It is interesting at this stage to note the ways in which various kinds

of structural apparatus aim to introduce numbers greater than 10. Cuisenaire, for instance, by including the 10 rod as the longest one available, emphasises that the orange rods (i.e. 10) should make up as much of the required length as possible, and then just one other rod is needed to complete the required length.
e.g.

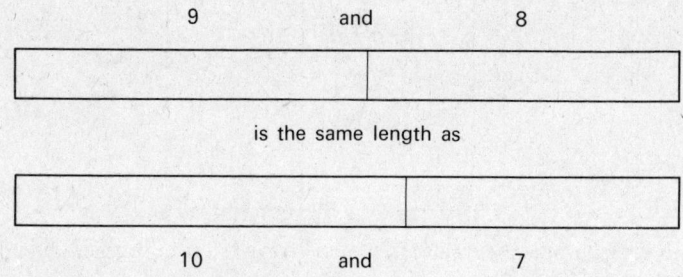

which we write as 17.

In the Stern equipment a 20-tray is included:

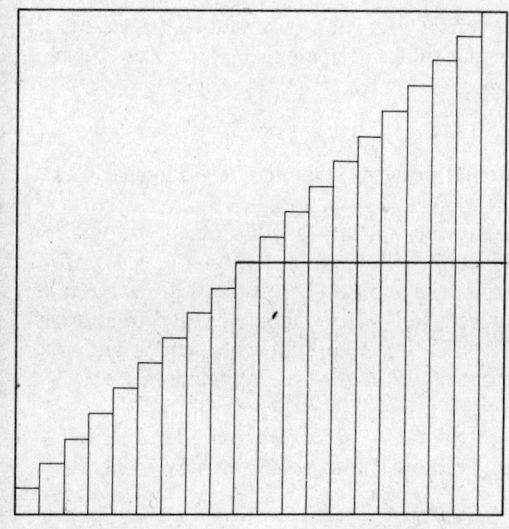

Dr Stern emphasises that the child should recognise that the rods used to represent 1 to 10 are repeated for the numbers 11 to 20, this time each one standing on a 10 rod.

Once the basic notion of place value is established most schemes tend to turn their attention to *addition bonds to 10*. Again, by using structural apparatus the child is given the opportunity to establish these.

68 Mathematics: Friend or Foe?

Using Cuisenaire apparatus, the 'story of 10' looks like this:

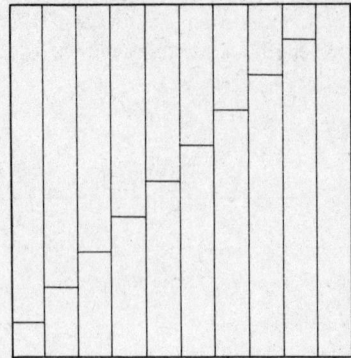

The Stern equipment supplies individual trays for each of the numbers up to and including 10.

These ideas lead naturally to addition and subtraction. As far as addition is concerned there is considerable support for *the counting-on principle*. Fletcher introduces counting on to 10 very early, leading to establishing addition facts to 5 and 6, then to 7 and 8 and then to 9 and 10.

The first ideas of *subtraction* vary from considering triples of numbers in which addition and subtraction are seen as inverses of each other, e.g.

$$5 + 3 = 8 \qquad 8 - 3 = 5$$

to comparison of (or difference between) two numbers. Fletcher uses the latter idea.

A more formal approach to the concepts of addition and subtraction is generally favoured at this stage with the introduction of *symbols for operations*. As children master the addition bonds to 10 this is the time when most schemes introduce traditional symbols +, − and =, although the → symbol also retains its use. Sealey uses the word 'equivalent' and the phrase 'is equal to' before the symbol =, while Fletcher uses the arrow, e.g. $10 - 6 \rightarrow$.

Continual recognition of the *relationship between addition and subtraction* is encouraged. Opinions about *recording* vary from no recording at all

to: ⟨7, 6⟩ ⟶ 13

to: (7, 6) ⟶ 13

to: ⟨7⟩ $\xrightarrow{+6}$ ⟨13⟩

to: $7 + 6 = 13$

to: $\begin{array}{r} 7 \\ +6 \\ \hline 13 \\ \hline \end{array}$

Knowledge of *numbers greater than 20* generally precedes the acquisition of *number bonds to 20*. The latter involves not only consideration of addition and subtraction (possibly in the form of counting on and counting back) but also the system of place value 'exchange' (exchanging 10 units for one 10) and the related concepts of multiplication and division, possibly in the form of products and inverses.

Most schemes stress the importance of the two aspects of *division*, although only brief attention is given to them at this stage of the work. Incidentally, considerable variation is introduced in connection with the vocabulary used in books and one wonders whether teachers and children are in danger of becoming trapped inside one particular system by specialist vocabulary - or alternatively whether they are excluded from joining the system later, for the same reason.

For instance, a child using Sealey's *Beginning Maths*, Book 3 (pages 8 and 9) would be familiar with this vocabulary:

Make this group:	We can put the things in *fives*.
Its number is 9	Like this
	There is *1 five* and *4 more*
	We can put the things in sixes, and sevens, and eights, as well. Do this. *The number of the group is always the same.*

One wonders whether the same child would be handicapped if he were transferred into a school using the Fletcher system in which he

70 Mathematics: Friend or Foe?

would find a different vocabulary being used:

> Partition each set into subsets with 4 objects in each.
>
> $\frac{8}{4}$ ☐ because ☐ (4) ⟶ 8

(Reproduced by permission from Fletcher: *Mathematics for Schools*, Level I, Book 6.)

By now, it is generally assumed that knowledge of *numbers in sequence to 100* can be consolidated in order that the foundations are firmly laid for further work on the four operations on number.

2.5 THE FOUR OPERATIONS ON NUMBER

We now consider the *four operations on number*, namely, addition, subtraction, multiplication and division. Fletcher concentrates first on *combining equivalent sets*, while Sealey groups in 2s, 3s and 4s. Other schemes move immediately to a more formal approach based first upon *addition facts to 10*, with use of the addition sign in recording.

The concept of addition is variously dealt with as:

(a) the union of disjoint sets;
(b) two lengths, end to end–the method generated by Stern or Cuisenaire structural apparatus;
(c) two steps taken in sequence, by counting on if required.

The concept of subtraction is established as:

(a) comparison of two numbers, i.e. the difference between two quantities, or
(b) counting back.

The desirability of relating the concepts of addition and subtraction at this stage is emphasised.

By this time, the *symbols* $+, -, <, >$ and $=$ are in common use, and the $\rightarrow$ is still favoured by some in recording. The combining and separating of sets and the partitioning of equal sets seek to reinforce the concepts of addition and subtraction while also paving the way for the later related concepts of multiplication and division. Fletcher makes much use of what he calls 'open sentences' involving addition and subtraction, in which the child deals mainly with sums which total to numbers between 4 and 12. It is interesting to note that at this point he introduces the idea of the identity element and zero in addition.

The commutativity of addition is stressed by most schemes, i.e. that $a + b = b + a$. Establishing *number bonds to 20,* and *counting on in 10s to 100* both involve grouping in 10s and further involve a basic use of place value in recording numbers.

The concept of place value is looked upon as an *exchange* or a *regrouping* of ten units for one ten, or vice versa. This terminology has replaced the traditional one of 'carrying to the next column'. Having used the concept of place value for two-digit numbers, *the numbers in sequence to 100* are reinforced. Since the increased and widespread use of metric measures the need for the understanding and use of higher numbers has already caused many teachers to reverse the order of their teaching and to concentrate on numbers up to 100 before spending time on number bonds to 20.

Basic ideas of multiplication and division are, in general, introduced before formal, more advanced work on addition and subtraction. *Multiplication* is seen as finding 'sets of' by Fletcher, while *Guidelines* favours considering multiplication as repeated addition.

In establishing the *concept of division, the two aspects of division* are introduced simultaneously in almost all systems. Normally these are termed 'grouping' and 'sharing', but they also have the names:

'measuring' and 'sharing'
'quotition' and 'partition'.

A simple explanation of these two aspects of division is given in The Schools Council's *Mathematics in Primary Schools* (page 23) and can be summarised diagrammatically:

	People	*Share*
Grouping	Number of people is *not* known	Size of share is known
Sharing	Number of people is known	Size of share is *not* known

72   *Mathematics: Friend or Foe?*

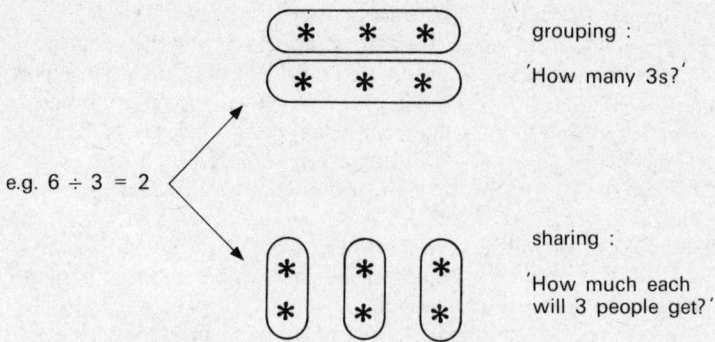

All of this work of combining together small numbers by the four operations is reinforced in various ways which include the use of the addition and subtraction squares and the process of counting on and back with such numbers as 10, 9, 5 and 3....

It is assumed that by now the child has a reasonable repertoire of facts about numbers and number bonds and of ways of recording these facts and so is ready to consider formal processes.

Fletcher deals with addition in four stages:

Tens to be added to tens
Tens and units to be added to units
Tens and units to be added to tens and units
Tens and units to be added to tens and units, with exchange.

Further use of the addition square is made to emphasise again the commutativity of addition and to establish also the associative law of addition, i.e. that if three numbers a, b and c are being combined, then

$$(a + b) + c = a + (b + c).$$

Subtraction is considered either as complementary addition, or as decomposition (or regrouping). In the particular schemes under review the process of equal addition is not used as a method to be taught.

EXAMPLE: 35−17
Complementary addition:
'Since 17 + 3 → 20,
  and 20 + 10 → 30,
   and 30 + 5 → 35
    then 3 + 10 + 5 = 18'

Decomposition

'$\begin{bmatrix} 35- \\ 17 \\ \rule{1cm}{0.4pt} \\ \rule{1cm}{0.4pt} \end{bmatrix}$ is the same as $\begin{bmatrix} 30 \text{ and } 5- \\ 10 \text{ and } 7 \\ \rule{2cm}{0.4pt} \end{bmatrix}$ which gives the same answer as $\begin{bmatrix} 20 \text{ and } 15- \\ 10 \text{ and } 7 \\ \rule{2cm}{0.4pt} \\ 10 \text{ and } 8 \end{bmatrix}$

18'

In view of the widespread use of subtraction by equal addition among the present adult population, it is perhaps surprising that it is not generally favoured at present as a method to be taught. Certainly for the purpose of doing mental calculation (which is probably the case for most of our numerical calculations anyway), subtraction by complementary addition is probably the safest method. It is the one favoured by shopkeepers who encourage their assistants to calculate the change required by a customer by adding on to the amount spent until the required total is reached, i.e.

Assistant: 'That will be 7p please.'
Customer: 'here is 10p.'
Assistant: (handing out pennies) '... 8 ... 9 ... 10p.'

So therefore 3 pennies are given as change.

As far as stages in subtraction are concerned, Fletcher's schemes again would seem to be favoured generally:

Tens and units ...   take away ...   units
Tens and units ...   take away ...   tens
Tens and units ...   take away ...   tens and units
Tens and units ...   take away ...   tens and units, requiring exchange

The form of exchange favoured by Fletcher is that of decomposition.
The operation of multiplication can be considered as:

(a) uniting a number of identical sets;
(b) putting end to end a number of equal rods; or
(c) taking a number of equal steps forward along the number track.

The use of the multiplication square (at least as far as 10×10) is advocated by most to emphasise the commutativity of multiplication, shown by the symmetry of the placing of numbers in the square. The value of the 100 square (consecutive numbers, ten in each row) is advocated for demonstration of number patterns in tables. As far as *multiplication tables* are concerned the 2 and 4 (and possibly 8) times tables are generally recognised as the first to be mastered. After that opinions differ about the order in which multiplication tables should be taken. An interesting progression is suggested in *Mathematics in Primary Schools*:

> First 2, 4, 8,
> then 3, 12, 6, 9,
> 5, 10,
> 7,
> 11.

On the other hand, Fletcher, after the 2, 4, and 8 tables, concentrates on the 6, 7, 9 tables.

The operation of *division* is usually accomplished in one of the following ways:

(a) by partitioning into groups of a given size;
(b) by repeated subtraction;
(c) by sharing among a given number of groups; or
(d) by considering division as the inverse process to multiplication (grouping related to multiplication).

The latter is of particular interest and importance when dealing with division by a fractional number and is dealt with on page 78.

Open sentences involving the use of all the four rules can provide some necessary testing and reinforcement at this point.

Place value to 999 is now tackled and this opens up the way for addition and subtraction of numbers of three digits and more scope for the demonstration of the *associative law for multiplication*, i.e. that:

$$(a \times b) \times c = a \times (b \times c).$$

A third law for numbers (the first two being the commutative law and the associative law) is involved now because it paves the way for long multiplication. This is the *distributive law* which enables us to break into manageable parts a calculation which involves numbers which are either outside our repertoire of number facts, or not number bonds provided by the normal multiplication tables. It is based upon the fact that:

$$a(b + c) = ab + ac.$$

For example:

$$\begin{aligned} 9 \times 26 = 9 \times (20 + 6) &= (9 \times 20) + (9 \times 6) \\ &= 180 \quad + 54 \\ &= 234 \end{aligned}$$

Recorded in the traditional way, and taking advantage of the commutative law that 9 × 26 gives the same answer as 26 × 9, this would be:

$$\begin{array}{r} 26 \times \\ \underline{9} \\ 180 \\ \underline{54} \\ \underline{234} \end{array}$$

In all cases, *long multiplication* follows the establishment of an understanding of the distributive law.

Long division follows hard on the heels of the long multiplication. Fletcher divides by numbers less than 10 to start with and then progresses to divisors less than 20.

Interspersed with the teaching of these essential processes, *Guidelines* introduce the basic ideas of *multibase arithmetic, indices, positive and negative numbers and modular arithmetic.* For enrichment purposes I favour these and any mathematical activities which, to quote Professor Matthews again, are 'pleasurable, profitable and pure'. Whether work is 'pleasurable' depends upon the skilful introduction of the subject by the teacher to the right children at the right time. With regard to the four ideas listed above, there is no doubt that they are mathematically 'pure'. So what about 'profitable'? I would define 'profitable' in this context to be either useful (in a utilitarian way) or mind-stretching. A case can be made for each of these extensions of number work mentioned above. For instance how 'profitable' is it for children to deal with multibase arithmetic? Perhaps they are

(a) using a method which applies universally to all measurement operations and conversions; this reason decreases in value with the phasing out of imperial measures; or
(b) manipulating with numbers in their own number system and hence gaining valuable practice and experience; or
(c) reinforcing their understanding of the structure of their own number system by having to consider the structure in another; or
(d) appreciating, in the binary system, the rudimentary principle of the two-state system required for processing numbers into a computer:

i.e. (0, 1) or (off, on).

2.6 FRACTIONS, DECIMAL FRACTIONS AND PERCENTAGES

Within their previous number work the children will have become familiar with facts about most of the numbers up to 100. The *factors* and *multiples* of a number will have been explored.

This is basic to all work dealing with *fractional numbers* which are written as the ratio between two numbers, i.e. rational numbers.

Words such as 'half' and 'quarter' are likely to have been used by the child from an early age. He may have a reasonable idea of the amount but no particular understanding of the concept. So after an oral experience of fractions, all schemes stress the importance of a good understanding of the *concept and meaning of a fraction* and of the *recording of fractions*. Fortunately, the inclusion of *realistic fractions* only is advocated - in particular halves, quarters, thirds, eighths and tenths.

Practical experience (with, for instance, structural apparatus, or an apple or a circle ...) is advocated by *Mathematics in Primary Schools* to establish the nature of a fraction. Pictorial representation of fractions is also recommended.

e.g.

 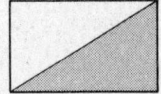 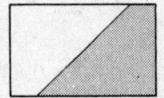 

In each case, half of the whole rectangle is shaded.

The *ordering of fractions* comes next. It is interesting to note that most schemes at an early point (with the notable exception of *Mathematics in Primary Schools*) advocate introducing the *meaning* and *recording of decimal fractions* as a way of recording tenths. In fact *Guidelines* deals with the idea of the decimal fraction before the vulgar fraction. However, most schemes agree that there is no distinct preference for *fractions preceding decimal fractions* although in most cases this is so. While metrication in forms of measurement highlights the importance of the decimal fraction, it is nevertheless true that virtually all necessary computation in the primary school can be accomplished using integers rather than decimal fractions, so that an early understanding of the decimal point is not essential, e.g.

2.65 metres × 4 can be calculated as 265 centimetres × 4.

In money calculations it is suggested that the point is temporarily explained as a barrier between the whole and fractional parts of the pound, e.g. £2.65.

Most schemes now feel ready to turn their attention towards operations on fractions and, apart from a preliminary canter doing the *addition of fractions less than 1, with the same denominator,* all schemes consider the concept of the *equivalence of fractions* of paramount importance as a prerequisite for the understanding of any formal operations. So Fletcher, for instance, introduces families of fractions, using denominators 2, 4, 6, 8, 10, for example.

The family of fractions equivalent to $\frac{1}{2} = \{\frac{1}{2}, \frac{2}{4}, \frac{3}{6}, \frac{4}{8}, \ldots\}$. When the child fully appreciates that there are many ways of expressing the same quantity, then he is ready, not only to *put in order a set of fractions*, but also to consider ways of combining two fractions using the four operations, in which so often it is expedient to express at least one of the fractions in an alternative way in order to do the combining, e.g.

> if $\frac{1}{2}$ is to be added to $\frac{1}{4}$, then the sum can be expressed as $\frac{2}{4}$ is to be added to $\frac{1}{4}$,
> i.e. $\frac{2}{4} + \frac{1}{4} = \frac{3}{4}$.

Addition and subtraction of fractions less than 1, with different denominators (where subtraction, for Fletcher, is considered as a process of comparison of size), is followed in general by the introduction of *mixed numbers* and an understanding of the *addition and subtraction of mixed numbers*.

Some schemes at this stage advocate a particular study of tenths, together with addition and subtraction (comparison) of tenths; then recording of addition and subtraction of hundredths. *Multiplication of fractions* is then considered in various ways. *Mathematics in Primary Schools* advocates practical experience in measuring, the Mathematical Association favours a diagrammatic explanation, Fletcher uses the word 'of' to start with. The formal stages of multiplication are usually:

(a) fraction × whole number;
(b) whole number 'lots of' a fraction;
(c) fraction × fraction;
(d) mixed number × mixed number.

The latter, in particular, can be made clear by diagrammatic representation:

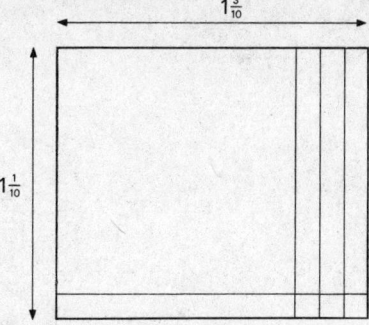

78 Mathematics: Friend or Foe?

With *division of fractions*, after a straightforward start by dividing a fraction by a whole number, all schemes deal with the traditionally troublesome *division of a fraction by a fraction*. *Guidelines* and Fletcher go immediately to the heart of the problem by introducing the idea of an *inverse* and the *identity element*.

Inverse fractions can be explained thus:

The (multiplicative) inverse of $\frac{3}{4}$ is $\frac{4}{3}$.
The (multiplicative) inverse of $\frac{4}{3}$ is $\frac{3}{4}$.

A fraction multiplied by its inverse gives the identity element.

i.e. $\frac{3}{4} \times \frac{4}{3} = 1$

So if our cause is furthered by eliminating the (multiplicative) effectiveness of a fraction by involving it with its inverse, then we may get nearer to solving our problem.

e.g. Problem: $\frac{2}{3} \div \frac{3}{4} = ?$

or put in another way

$$\frac{2}{3} = ? \times \frac{3}{4}$$
$$\text{Now } \frac{2}{3} \times \frac{4}{3} = ? \times \frac{3}{4} \times \frac{4}{3}$$
$$\text{i.e. } \frac{2}{3} \times \frac{4}{3} = ? \times 1$$
$$\frac{2}{3} \times \frac{4}{3} = \boxed{\frac{8}{9}}$$

However, when teaching *division by fractions*, many teachers resort (perhaps wisely) to imposing a technique which works, namely, 'turn the fraction upside down and then multiply'.

The Fletcher books include a useful discussion of *three methods* which can be used to explain this process (*Level II Teacher's Resource Book*, for Books 7 and 8, page 31). All these three methods are then used one after another in the children's book (*Level II*, Book 7, pages 24 and 25). I believe that this multiplicity of methods leads to confusion in the children's minds and that the teacher should select whichever *one* method is most suited to the children at the time.

There are, of course, other methods available.

(a) A demonstration with a diagram: $3 \div \frac{1}{4}$

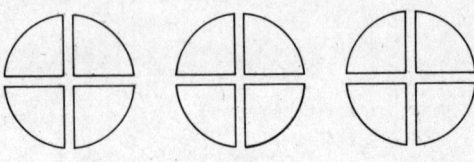

Answer: $3 \div \frac{1}{4} = 12$

$\begin{cases} 3\frac{1}{4} \div \frac{3}{4} \\ \text{How many three-quarters are there in } 3\frac{1}{4}? \end{cases}$

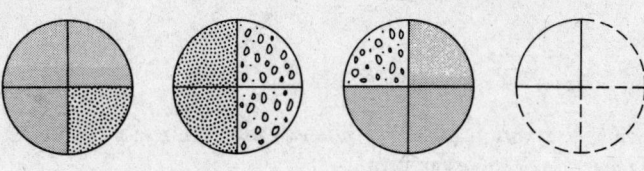

Answer: $3\frac{1}{4} \div \frac{3}{4} = 4\frac{1}{3}$

(b) Using the principle of a common denominator (in reality Fletcher's method 1, but recorded differently):

$$3\frac{1}{2} \div \frac{2}{3}$$
$$= \frac{7}{2} \div \frac{2}{3}$$
$$= \frac{21}{6} \div \frac{4}{6}$$

Since these are all sixths, then this is equivalent to

$$21 \div 4$$
$$= 5\frac{1}{4}.$$

For those schemes which have not been keeping work on fractions and decimal fractions parallel with each other, it is stressed that *decimal notation* uses the concept of place value to record tenths, hundredths. In *Mathematics in Primary Schools* the *four operations on decimal fractions* are made to arise from practical problems. When place value is thoroughly understood, *addition and subtraction of decimals* follows readily.

Children are encouraged when multiplying by a decimal fraction (a) not to think in terms of moving a decimal point (it is not the *point* which moves but the numbers which, by increasing tenfold or a hundredfold, change to a different column), and (b) to reason beforehand the approximate size of their answer in order to check whether their answer digits are placed correctly about the decimal point.

With *division by decimal fractions* it is advocated that this be done either

(a) by converting the decimal fraction into an 'ordinary' fraction and using a previously used method, or
(b) by using the principle of multiplying the sum by the identity element, expressed in whichever of the following forms will make the division into a whole number:

(all equal to 1): $\frac{10}{10}, \frac{100}{100}, \frac{1000}{1000} \ldots$

e.g.
$$\frac{6.3}{0.7}$$
$$= \frac{6.3}{0.7} \times \frac{10}{10}$$
$$= \frac{63}{7}$$
$$= 9.$$

Then follow various brief introductions to such related topics as *ratio, recurring decimals* and *scale*.

It is advocated by *Mathematics in Primary Schools* that the first encounter with *percentages* should be practical, in which the children meet a new small unit of $\frac{1}{100}$ or %, this being a convenient form to use when comparing or using fractions, whether in vulgar or decimal form.

2.7 MEASUREMENT

2.7.1 *Length, weight (mass) and capacity*

Before dealing specifically with these forms of measurement it is necessary to outline some basic principles which underlie all the schemes under consideration:

(i) Fundamental and derived units of measurement. The subject of measurement involves many different *fundamental* forms:

>Measurement of length
>>weight (mass)
>>capacity
>>time
>>temperature
>>angle
>>money

and also some *derived* forms:

>Measurement of area
>>volume
>>density

(ii) Vertical and horizontal approaches to measurement. It is generally considered necessary, by each of the schemes considered here, for both the vertical approach within one type of measurement (e.g. length) and the horizontal approach across several types of measurement (e.g. length, weight, capacity) to be investigated.

The vertical approach of these schemes is analysed briefly in the following section and it will be seen that progress through work with certain of the measures runs parallel with others and that therefore horizontal links are generated.

For instance, the routes through work on *length, weight (mass) and capacity* proved to be so similar that one summary of it suffices for all three.

Similarly the approach to work on area and volume is shown to follow roughly parallel routes and again horizontal links are forged. Since the units of measurement used for area and volume are derived from the fundamental unit of length, it is appropriate to consider at the same time, progress in *length, area and volume*.

Volume and capacity are considered by some to be synonymous. The actual link between the two is aptly described in *Metric Units in Primary Schools*:[1]

'Boxes holding just 1000 cubic centimetres, whatever their shape, have capacity one litre. This gives a useful link between the metric unit of capacity, the litre, and volumes which are derived from lengths measured in centimetres, whether the volume is space taken up by a solid or a liquid, a displacement, or space within a hollow vessel.'

When using imperial measures we have been able to recall delightful mnemonics such as: 'A pint of water weighs a pound and a quarter' and 'A heaped tablespoonful of flour weighs an ounce' How simple it will be to use the horizontal link between *volume, capacity and weight (mass)*.

$$1 \text{cm}^3 \equiv 1 \text{ml} \equiv 1 \text{g (of water)}$$

Having established a basic *vocabulary* for measurement, the *concept of measurement* is introduced as a *form of comparison* between one length and another, between one mass and another, between one container and another. Lengths, masses and containers are *put in order of 'size'* but as yet no attempt is made to quantify the item being considered. It is important that *conservation* of length or of mass or of capacity is appreciated and all practical work at this stage will aim to establish these basic ideas.

Comparisons made will be estimated by the child who may devise his or her own criteria for making an opinion, e.g. 'This pencil is shorter than that one because I can hide this one in my hand' or 'This box is heavier than that one because I can pick that one up but this one is too heavy'. The *idea of balance* and of *equality* is an important one to be grasped as it is the basis upon which all measurement is made, e.g. this stick is as long as six squares on this paper.

To establish further the concept of measurement as a form of comparison, each scheme favours the *use of arbitrary units*, invented by both teacher and child. However, included in these arbitrary units may well be the standard units which will be established later. The Mathematical Association report points out that 'the simplification brought about by developing the measurement of length, area and volume from apparatus already used as an admirable representation of number (Cuisenaire, Colour-factor) will give much greater surety to the understanding of measuring'. Fletcher includes the metre among the arbitrary measures used when measurement is first introduced in *Level I*, Book 3.

It is worth noting, perhaps, that Fletcher introduces very little work on measurement within *Level I*, but that he does give the child a brief contact with the concepts of length, capacity, mass, time and volume.

There comes a time when arbitrary measures become inadequate, as different children find they get differing answers to the same measurement task, e.g. when comparing the circumference of a football with their span. The introduction of *standard measures* becomes necessary and appropriate. Certain difficulties are being experienced by teachers due to the fact that some *metric units* are either too small or too large for teaching purposes with the result that (a) children are needing to use a large number to express a comparatively small measurement before they fully understand the large number, and (b) fractional parts of the large standard unit may be needed, since the rounding off to a whole number introduces too great an error.

An example of (a) is that of an object of such modest size as a small tin of baked beans, which weighs 238 grams. An example of (b) arises when a child may be finding the length of a room which is, say, 8 metres 75 centimetres long. Which is the best form of answer, initially, for that child to use?'

8m 75cm
or 8m
or 9m
or 875cm...?

Teachers are devising their own means of avoiding these difficulties. For instance, one teacher of very young children uses the word 'hundredgram' as a unit so that the tin of beans weighs just over 2 hundredgrams. In length, the Cuisenaire 'ten' rod, or the decimetre, becomes a unit of length for some.

It is necessary at this point to discuss briefly the relationship between *mass and weight.* It is interesting to note that both *Guidelines in School Mathematics* and the Mathematical Association report steadfastly use the word weight, and not mass, in connection with the activities done in the primary school. In contrast to this, Fletcher and the Royal Society booklet *Metric Units in Primary Schools* stress use of the word 'mass', in view of the fact that weightlessness is now a well-known phenomenon.

The layman's view of the relationship between mass and weight might be as follows. All objects are composed of matter. Matter is the 'stuff of the universe'. The mass of a body can be defined as the quantity of matter in it. Any two objects exert a mutual gravitational force of attraction on each other but for objects of ordinary everyday sizes this force is extremely small and difficult to detect. If, however, one or both of the objects becomes very large, for example, if one object is the Earth and the other object is a potato, then the mutual force of attraction is easily detected. Although the force of attraction that the potato exerts on the Earth is equal to the force the Earth exerts on the potato, we are normally only aware of the latter and we call this force the 'weight' of the potato. (Query: How could we compare masses in free space?)

However, perhaps the ATCDE (*Children Using Mathematics*) should have the last word:

'We propose that no great fuss should be made about the distinction. The word "weight" should be used according to common custom, although we feel that the teacher can use the word "mass" where he knows it to be more appropriate, so that the children at least meet the word in correct context.'

The ATCDE make an interesting subdivision of standard units, suggesting that *appropriate teaching units* may well be:

Length	$\frac{1}{2}$m $\frac{1}{4}$m m
	cm
	km $\frac{1}{2}$ km
	mm
Weight	kg $\frac{1}{2}$kg $\frac{1}{4}$kg
	200g 100g 50g 20g
Capacity	$\frac{1}{2}$l $\frac{1}{4}$l

On the other hand Fletcher advocates getting children to use the metre first, then the centimetre, and, for weighing activities, the 100g and 10g weights. The Worcestershire scheme suggests yet another set of length measures for teaching purposes, the centimetre first, then the metre, then the millimetre, then the kilometre.

The *relationship between units* within any one form of measurement needs to be established as this leads to *computation using units of measurement*. The ACTDE outlines some basic computational skills which are required for measurement and calculation. These include addition and subtraction of fractions (particularly $\frac{1}{2}$ and $\frac{1}{4}$), addition and subtraction of decimals (to two decimal places), multiplication of and by fractions, and division by integers. The Mathematical Association includes $\frac{1}{3}$ and $\frac{1}{6}$ among the fractions and also emphasises the duality of multiplication and division in computation.

The various *processes used in taking measurements* need to be understood and practised, with the *use of instruments* which are appropriate to the type of measurement, appropriate to the degree of accuracy required and appropriate to the skill of the child doing the measuring. *Estimation* is considered by all schemes to be an important preliminary activity to the actual task of measuring, and also consideration of what constitutes an appropriate *degree of accuracy*, and what amount of *error* can be tolerated, e.g. for travel purposes, it might be sufficient to know that a certain bridge is 'a quarter of a mile long' but for construction purposes the same length will need to be given precisely.

Guidelines lists a number of *applications* of these forms of measurement and include *shopping, force of gravity, plans and scale drawing, the circle, speed.*

The inter-relationship of different types of measures are dealt with elsewhere (see pages 28, 89).

2.7.2 Length, area and volume

Handling and observing objects is the starting point suggested for work on length, area and volume. The child will be asked to make a *comparison* of 'size' and to sort items into an order of size according to his own criteria and using his own invented methods.

It is now necessary to consider in general the nature of *units of measurement* which will be used and, in particular, for the teacher to appreciate that, while measurement of length uses a *fundamental unit*, area and volume require *units derived* from considering the length in two and in three dimensions respectively. This may well be the first time that the child has met the need to express a quantity using a product of two distinct (or of three distinct) parts and so the possible difficulty must not be underestimated. (Another example in which the child may have met the same idea is in fraction work where one quantity is expressed as the ratio between two numbers, e.g. $\frac{3}{4}$.)

And now the teacher faces the question: *'Which should be dealt with first? Length? Area? Volume?'*

Opinions differ from scheme to scheme. The ATCDE document

maintains that length, area and volume 'are all three abstracted from the concept of extension in space' and that *length* 'is a convenient starting point for quantitative treatment of measure'. The further point is stressed that, of the three measures, length is the simplest one with which to form comparisons and to estimate equalities and inequalities.

Most schemes agree that linear measure should come first but that, since three-dimensional objects are those with which the young child is more familiar, then the concept of volume is more readily understood than that of area. However, while initial ideas of area involve only a comparatively simple square-counting process, the practical difficulties of cube-counting to find volume are enormous in comparison and are sufficient to delay work on volume.

Whichever measurement comes first, the four concepts of finding *the most suitable unit* to use as a measure (a square for area, a cube for volume), using the unit in this shape to *assess the area and volume* (by counting), appreciating *conservation of area and volume,* and thus appreciating *methods* of finding different shaped areas and volume, provide the essential core of the work.

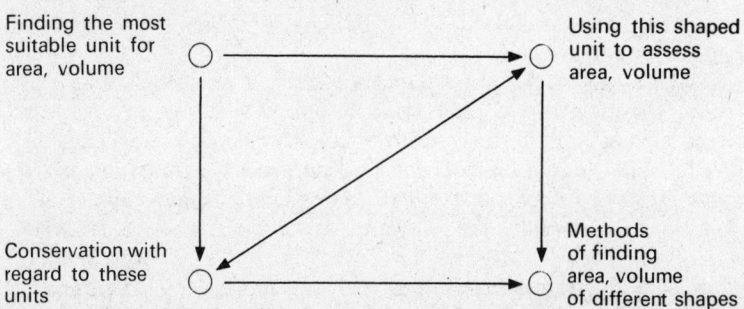

As teaching units, the centimetre (cm), the square centimetre (cm^2) and the cubic centimetre (cm^3) find most favour. It must be remembered, of course, that the centimetre is not an officially recognised SI unit but its usefulness as a manageable unit to be used by children must be accepted. Once the concept of measuring length is established with children and the need for finer measures arises, then the range of SI units of length may assume their greater importance. A practical difficulty which arises later is the use of large numbers associated with the measurement of the volume, in cubic centimetres, of objects of only modest size. The cubic decimetre (1000 cm^3) is useful when comparison with a litre is made. Fletcher follows the use of cm^3 with mm^3, while the ATCDE book goes to the larger unit, namely the cubic metre (m^3), stressing the importance of the conservation principle of considering m³ as a cube and as a cuboid. In area work, too, the ATCDE go for cm² and then m².

Initially, having established the use of standard units, *counting squares* provides a method for arriving at areas. A reasonably close approximation is achieved within an irregular figure by including in the count all complete squares and those squares for which more than half is within the area and by excluding all other parts of squares. The counting-squares method also has the advantage of leading directly to establishing the fact that the area of a rectangle is given by the product of the number of squares in each row (the required number is given by the length) and the number of rows (the required number is given by the breadth).

i.e. area of the rectangle = length × breadth

The relationship between *area and perimeter* is explored now. Surprisingly, perhaps, most schemes consider these concepts simultaneously, presuming that the children will have such a grasp of the basic essentials of linear and square measure that they will not be tempted to confuse the concepts of perimeter and area!

Calculations using standard units for area and volume problems require also a firm knowledge of the *relationships between units* within one form of measurement.

The *areas of shapes other than the rectangle* now need to be considered and after the child has used some *ad hoc*, invented methods for arriving at a result, the concept of conservation of area will enable him to relate the *area of a triangle, a parallelogram, a trapezium and a circle to that of a rectangle* and from this relationship to appreciate that *the basic formulae for calculating the areas* can be derived (see page 24).

Calculations in volume are achieved in a similar way. The basic formula for the *volume of a cuboid* is seen as the product of the number of cubic units required for the bottom layer (the number given by the length and breadth of the base) and the number of layers (the required number being given by the height of the cuboid).

i.e. volume of cuboid = length × breadth × height
= area of base × height

Following on from this, the *volume of any prism* (any three-dimensional block which has at least one uniform cross-section) is seen similarly as being derived from the product of the number of units of volume required for a bottom layer (which number is given by the area of the uniform cross-section, as the base) and the number of layers (number given by the height).

i.e. volume of a prism = area of base (cross-section) × height

The *relationship between the volumes of cuboids and pyramids; cylinders and cones* (pyramids on a circular base); *cubes, cylinders, spheres and cones* is found in a practical way first and then the *formulae* are derived from previous formulae established.

Extensions of this work on area, volume and perimeter arise through the relationships between area and perimeter, and between volume and surface area. These provide a rich field of practical and simple theoretical work and most schemes attempt to exploit these connections.

Similar shapes provide a useful link with number work (sequences) and graphical work.

2.7.3   Volume and capacity

Very little work is included on *capacity* throughout these schemes. It is practical and experimental and it concentrates on establishing the concept of the size of the units used and not on any accompanying calculation.

For instance, after brief preliminary work using cupfuls of water or sand to mark off graduations on a measuring jar (*Level I*, Book 2, pages 16 and 17), Fletcher proceeds immediately to introduce only one activity, using 1 litre and 2 litres. Apart from this, further work on capacity is delayed until *Level II*, Book 8 and then it is only to find the relationship between millilitres and cm^3 so that the volume of a cylinder can be approached via its capacity.

For convenience, the litre is at first broken down into the $\frac{1}{2}$ litre and $\frac{1}{4}$ litre, and eventually to the millilitre.

Through *play with objects and containers* the first experiences of the child involve *comparison of 'size'* and sorting into 'sizes'. It is important that the concept of *conservation* should be understood and this can be achieved through play with apparatus such as plasticine and individual wooden cubes (which can be rearranged into a variety of overall shapes) for the work on capacity. It is advocated that putting objects into *order of 'size'* is best done by comparing only two objects to start with and then later giving the child more than two objects to put in order.

Arbitrary units of measure are used to compare the capacities or the volumes of two hollow containers. These units may, perhaps, be marbles or eggcupsful of water, or wooden cubes. When volume is being calculated by this counting process, discussion about the suitability of the arbitrary unit as a space-filling one will lead to appreciation of the need for a standard unit which fills up the whole space. Also other ways of measuring capacities may well be devised, such as pouring water from a container with a known volume into the required one. In this respect *Guidelines* advocates the *construction of a measuring jar*.

This is a useful means of reinforcing (a) the fact that measurement is a form of comparison, (b) the concept of the size of the units, and (c) the concept of conservation of the unit.

ATCDE suggest that, in capacity, the teaching units should be litres, $\frac{1}{2}$ litres and $\frac{1}{4}$ litres, while the county scheme uses litres and millilitres.

Having established the *relationships between the units within one form of measurement*, most schemes point out the *relationships across forms of measurement*.

$$\text{i.e. } 1 \text{ litre} : 1000 \text{ cm}^3$$
$$1 \text{ ml} \ \ : 1 \text{ cm}^3$$

It is interesting to note that the ATCDE suggest that children should meet the cubic metre (m^3) as a cube and as a cuboid.

Calculations involved in volume and capacity work will be limited. In some schemes computation concerned with volume is restricted to the rectangular block. In capacity computation will involve the operations of *addition, difference, sharing* and *grouping*.

An interesting link between volume and number is provided by a consideration of *similar shapes*.

How volumes grow

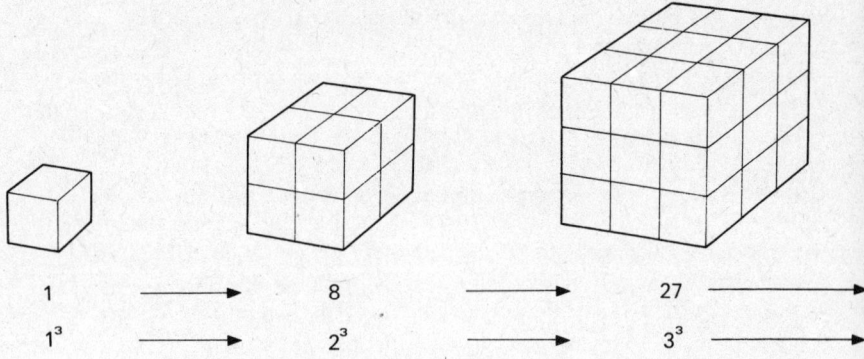

Density requires a little discussion here. In most cases it is added as an optional afterthought and there seems to be considerable reluctance to elaborate. Perhaps this reluctance is justified and is telling us that the concept of density is a difficult one and better omitted from the primary school classroom. However, the teacher will not be able to avoid for ever children's questions about why some big things float and why some small things do not and why two parcels, which look identical, weigh differently from each other.

The Mathematical Association report has a useful section on volume, weight and density (page 120) and describes practical experiments on floating and sinking carried out by children 'leading to the idea of density', but comments that '... any further pursuit of the idea of density seemed beyond them'.

The ATCDE book similarly makes small mention of density. At the end of a section on capacity, volume and area it includes, rather vaguely, as a progression in the work, the words 'possible link with density', but it does not attempt to enlarge on the subject.

When one realises that the units which children are likely to meet when density is recorded will be grams per cubic centimetre (g/cm^3 or $g\ cm^{-3}$), one can appreciate that the primary school contribution to this work is better restricted to the purely experimental and practical approach with objects of differing textures, shape, size and weight (mass). The standard form is kg/m^3 so if we record the density of water (which is 1 g/cm^3) in standard form, the numeric value is 1 000 kg/m^3.

Volume, weight (mass) and capacity. Most schemes emphasise the useful link which exists between these three forms of measurement when considered in metric units:

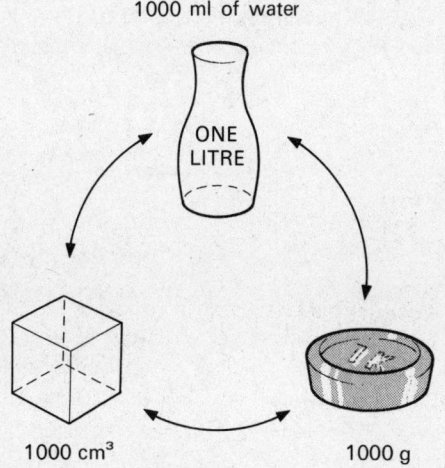

2.7.4 Angles

The importance of introducing first *the dynamic aspect of angle, rather than the static*, is emphasised throughout. A method which has support is that of children experiencing a full turn, $\frac{1}{2}$ turn and $\frac{1}{4}$ turn. This leads to a more specific consideration of the *right angle*. Fletcher

90 Mathematics: Friend or Foe?

considers the right angle in two ways - first as a fixed difference between two directions and second as a measure of rotation.

There is considerable difference of opinion about the appropriate time for measurement in *degrees* to be introduced. While ATCDE discuss the use now of a circular protractor with which the term 'turning' refers to the dynamic aspect of angle and 'corner' refers to the static aspect, a county scheme mentions using protractors with 5° markings. Fletcher and *Mathematics in Primary Schools* delay mention of degrees for some while, preferring to introduce the eight cardinal points in *compass bearings* to reinforce the concept of angle. There is variation of opinion about when these should be introduced. Most schemes favour using them early in the programme, four or eight points only, and not connected with the degree measurement.

The terms *horizontal and vertical* are used at this point.

Symmetry is recognised as a most useful mathematical phenomenon which gives rise to valuable work on the properties of shapes, and in particular to the angular properties which are highlighted by rotational symmetry.

The two main aspects of symmetry investigated at this stage are bilateral and rotational symmetry. It is noted that shapes which have bilateral symmetry about two axes of symmetry which are at right angles to each other automatically have rotational symmetry, e.g. the letter H.

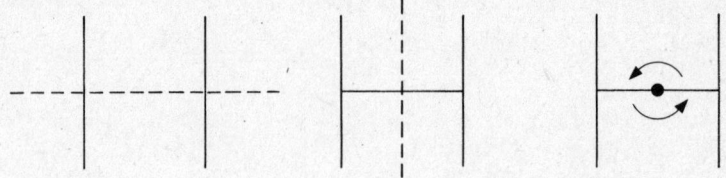

Tessellating shapes provide not only an attractive activity but also a way of finding out about the angular properties of shapes which either do or do not tessellate. For instance all regular polygons in which the size of the interior angle is a factor of 360 will tesselate:

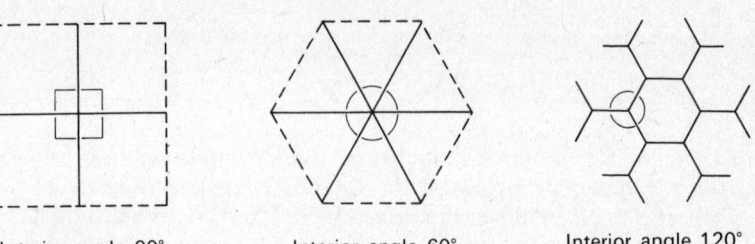

Interior angle 90°   Interior angle 60°   Interior angle 120°

Specific study of *the right-angled triangle* leads on to the discovery that *the sum of its angles adds up to 180°*. A typical method to demonstrate this is:

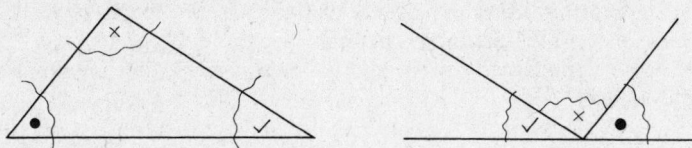

The *angle properties of any triangle and all quadrilaterals* are considered next and this leads in many schemes to the need for the *measurement of angles* and the *use of the protractor*. Fletcher's method of introducing the concept of measuring an angle is to compare one angle (called a unit angle and which has been cut out of paper) with a right angle and to record how many times the unit angle will fit into the other (*Level II*, Book 3, pages 53, 55, 56). This links, rather satisfactorily, the static and dynamic aspects of angles.

The circular protractor, divided into 10° sections, is advocated as the best instrument to use, since previous work on angles has been based on the full turn.

Considerable practice with measurement of angles is necessary for a certain acceptable *level of accuracy* to be achieved. Fletcher advocates *estimation* first followed by measurement.

The *use of compasses* in the *construction of angles* leads on to work connected with the *angles in a circle*, subtended by a chord to the centre and to the circumference.

Extensions of work on angles leads to such topics as *transformations* (in particular rotations about a point) and *vectors*. These are now an established extension of graphical and angular work in the primary school. They are used:

(a) to record a movement in two-dimensional space, noting both its magnitude and direction; and
(b) to visualize the effect of two (or more) consecutive moves. Vector addition gives a means of treating a set of numbers as a single entity.

A useful description of vector work is given in the Mathematical Association report (pages 102 and 103).

2.7.5 *Time*

There is almost total agreement on the *progression of work* in the primary school on time.

92 Mathematics: Friend or Foe?

The ATCDE book puts strongly what other schemes imply – namely, that no computation is needed with this work (page 86):

'So far we have not found any common uses of clock time (as distinct from elapsed time) that require calculations. . . . We feel that, if children learn to tell the time – and many do this at home anyway – and become familiar with these day-by-day situations, they are doing all that is needed.'

Having met the *basic vocabulary* related to time the idea of *sequence of time* is emphasised. While *Guidelines* uses words such as 'before, after, daily . . .', other schemes concentrate on events of the past, the present and the future. The child then learns of *the special times* during the day and the week as shown on the clock face. In some schemes the *calendar* is considered here, the child considering the *day*, the *year*, the *seasons* and the *moon*. Methods which have been used in the past for telling the time, e.g. sand clocks, candle clocks and water clocks, are sometimes constructed by, and with, the children to indicate *methods by which the passage of time can be measured*. This provides a foundation on which the teaching of telling the time is built. Fletcher, in fact, introduces the idea of a *time interval* as his first mention of time activities (*Level I*, Book 6). He does not deal with telling the time until *Level II*, Book 1. *Telling the time to the nearest* hour, quarter hour, 5 minutes and 1 minute is a generally accepted progression. This is followed by the *recording of the time and the date. Measuring time in seconds and in minutes*, sometimes by means of a seconds pendulum, and other times by the use of a stop watch to measure time in seconds and in minutes, follows. *Timing of activities* using a stop watch is included in every scheme and activities suggested range from timing fifty skips to timing reaction times.

Work on the *24-hour clock* is included in every scheme followed by consideration of *a.m.* and *p.m.* Thereafter the use of *timetables* provides the grand finale of this section of the work.

I am not sure whether the stress on *timetables* in each scheme is in order to ensure that the children can actually read a timetable, or whether it is in an effort to find a useful outlet for work on time. However, the activity is certainly a useful way of evaluating the child's grasp of methods of recording time!

2.7.6 Money

Decimal currency has supplied a most useful application for work on number and is now seen in this light. Fletcher for instance, readily calls on an example to do with money whenever he wishes to clinch yet another step in number – addition, subtraction, sharing and multiplication, decimal notation, long multiplication, fractions, decimal fractions.

The decimal point need not cause difficulties (see note in 2.6) if it is considered 'as a form of punctuation to separate complete £s from the fractional part' (Mathematical Association). The method suggested by the ATCDE is to 'think of the decimal point as a separator between the £ and the pence'. This seems to me to be an unfortunate way of expressing why the decimal point is there and may even help to perpetuate the widespread error in recording prices like this: £2.50p! (2.50 is *one* quantity according to our place-value system, so how can it be '£' and 'p' at the same time?)

Guidelines provides the child with *exchange play* using a form of *shopping* before starting where other schemes start, namely at *coin recognition*. When both concepts (of exchange and of coin recognition) are mastered the child is ready to be involved, through 'shopping', in the *more formal operations*. Initially these are addition and multiplication, giving totals less than 10p, followed by subtraction as complementary addition and, later, both the sharing and grouping aspects of division of money.

The *relationship between units of money*, together with use of the £, p notation, leads on to *operations in decimal form* where necessary. In particular one scheme stressed the order:

$$£, p \times 1 \text{ digit}$$
$$£, p \div 1 \text{ digit}$$
$$\text{Fractions of £1.}$$

Use of *a ready reckoner* is taught and encouraged and Fletcher introduces the idea of other currencies and *foreign exchange* as a means of applying the concepts and techniques introduced.

2.8 SHAPES

The study of shapes is so diverse that it is small wonder and totally acceptable that there is no clear-cut syllabus through it. However, it is important that shapes activities should not appear to be a number of isolated artistic activities but that one skill acquired, or piece of knowledge learned, can be used or applied for the next activity. Handling of *three-dimensional objects* seems to be a generally accepted start and the differences of opinion mainly arise over the *vocabulary* to introduce. While Fletcher introduces names such as cube, cuboid, cylinder, sphere and prism right from the start, other schemes are content to deal with 'balls' and 'blocks'. While it seems unfortunate for the teacher not to use the proper name for a shape to a child, it seems equally unfortunate to require a child to be restricted to the one official name and to forget the descriptive label which he may have acquired previously, such as ball, round, ring, oblong, diamond, block, brick.

Handling of objects in the environment gives rise to a consideration of certain *three-dimensional objects* and the recognition of the *cube, cuboid, cone, cylinder and sphere*. Investigating the *general properties of three-dimensional objects*, e.g. fabric, 'size', feel, hardness ... by activities involving sorting and classification will serve to establish the *vocabulary* involving the names of the shapes and such words as line, face, point, angle. *Two-dimensional shapes* will need to be recognised in connection with three-dimensional objects and the ones which usually come first are, of course, the square, rectangle, triangle and circle.

Consideration of the faces of an object leads to such activities as surface covering, making *patterns* and *tessellations*. For mathematical activities (such as making patterns and tessellations) with the basic 2D shapes (square, rectangle, triangle, circle) it is generally recommended that the young child be supplied with the shapes to use, or *templates* to draw round or with strips with which to construct shapes, as the skill to make shapes which are sufficiently accurate will not come until later on.

The *2D and 3D approach* to shapes can become very arid. Perhaps other classifications which may prevent the customary 2D and 3D categories from being set could be:

>   natural shapes/man-made shapes
>   regular shapes/irregular shapes
>   symmetrical shapes/non-symmetrical shapes
>   permanent shapes/shapes generated by movement
>   (e.g. radiating concentric
>   circles caused by tossing a
>   stone into still water)

Close observation of these shapes will help the child to make comparisons, to notice shapes in the environment and *to make shapes* from given pieces for use in patterns.

The *uses of shapes* will be reinforced by looking at composite shapes, by classifying shapes, by finding *connections between shapes and number*, by investigating the *rigidity* of shapes and their *symmetry*. By studying the *symmetry* of a shape each scheme recognises that the properties of the shape become more apparent. For instance, with 2D shapes the properties can be defined as the nature of the sides, angles and diagonals, while with 3D shapes the properties are defined as the nature of the faces, the edges and vertices. The main forms of

The Syllabus

symmetry emphasised in these schemes are *bilateral and rotational symmetry*.

Devising suitable *nets of polyhedra* is an activity which is useful both mathematically and artistically as it leads to serious consideration of the properties of the shapes being considered. Games requiring dominoes, pentominoes or hexominoes are useful here.

Fletcher follows work on symmetry by looking particularly at the *properties of the square, rectangle, parallelogram, rhombus, isosceles triangle and equilateral triangle* and a little later he deals with the *properties of the circle* including the relationship between the diameter, the circumference and π.

Transformations and invariance are encountered next. The four kinds of transformations introduced into all the schemes are: *reflection of a shape, rotation, translation* and *enlargement*. The Mathematical Association report gives an authoritative introduction to this work (pages 101–9).

Knowledge of the *geometrical properties of shapes* is continually being reinforced and in particular two outcomes are emphasised:

(a) the inter-relationship of the number of faces, edges and vertices of platonic solids;
(b) Pythagoras' principle.

There is a natural overlap between work on shapes and work on area. All consideration of angles, perimeter, area and volume is facilitated by general experience with shapes as outlined above.

Some *extensions* of this work on shapes may be on *vectors* (see 2.7.4), *similarity and enlargement*. Simple work on some *conic sections* may follow, in which the circle, the parabola and the ellipse are considered, either as lines made up of an infinite number of points, for which the co-ordinates of each of these points illustrate a certain relationship between them, or alternatively as the path of a moving point which obeys certain restrictions as it travels.

2.9 GRAPHICAL REPRESENTATION

Most schemes recognise that graphical work is not a subject in its own right but that it supplies us with *a means of recording data* which is both visually appealing and readily interpreted.

The scheme which does not use a graphical method to record data early in its programme is Fletcher, apart from brief tabulation of numbers in *Level I*. A bar chart is used briefly in *Level II*, Book 1, to be followed very soon afterwards with the order pair concept. This approach differs considerably from the traditional approach.

Most schemes start with having a number of objects for *counting* and for *comparison*. A *one-to-one correspondence* is set up between one piece of data and one block and then the block is used:

(a) in a *comparison of two categories* only (e.g. those children who stay at school for dinner and those who do not);
(b) in a *comparison of more than two categories*.

The *arrowgraph* is another simple form of pictorial representation which is used at this stage by young children to record a relationship which exists or which has been investigated. The arrowgraph (sometimes called an arrow diagram or an arrowgram) is useful initially in that young children have sufficient skill to make their own recording but the usefulness is quickly dissipated when too many arrows on the diagram cause confusion and the data cannot be retrieved easily from the graph.

Next the *pictogram* is used by some schemes. One type of pictogram is a graph in which a number of identical pictures or symbols are used in columns or rows, each symbol representing a given quantity. The other method of making a pictogram is by using symbols which are enlarged according to the frequency of the data. Errors, deliberate or otherwise, can occur in these due to the fact that the reader needs to know whether the enlargement factor has been used in a linear form, or area form, or volume form, e.g.

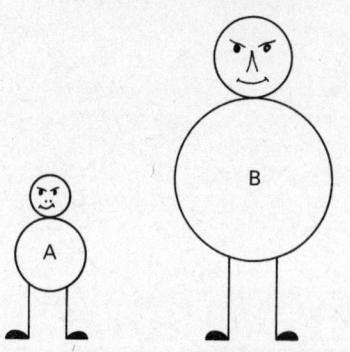

Does B represent *twice* as much as A ?
(linear comparison)

or

Does B represent *four* times as much as A ?
(area comparison)

or

Does B represent *eight* times as much as A
(volume comparison)

The most popular form of graph used in the classroom is undoubtedly *the block* (or *column*) *graph*.

A mistaken view is that *column graphs* are suitable for young children and that when one matures one creates line graphs instead. All schemes attempt to stress that it is the *data* which determine the type of graph used and not the age of the graph maker. Nevertheless, since a column graph is simpler to make and to interpret, the teacher of young children needs to use situations such that a column graph is the best method of recording the data. For example:

Simple graph projects resulting in block graphs
(a) One-to-one correspondence -
    favourite ... (animal, food, colour),
    types of ... (shoe, dinner).
(b) Where each child has a column,
    e.g. number of boys and girls in the family.

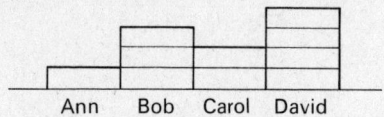

(c) number of pathways (see page 40),
    box.

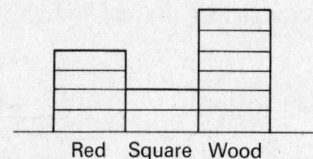

(d) Census - traffic, birds, animals.
(e) Dice - addition, subtraction, multiplication of the numbers shown.

Block graphs revealing a pattern
(a) multiplication tables,
(b) frequency of shoe sizes,
(c) number of pathways (see page 40),
(d) targets,
(e) dice scores.

The introduction of the idea of a *scale* on the axes follows naturally from experience with block graphs. The numbers on the vertical axis are introduced to facilitate the reading of the frequency of data since it will no longer be necessary to count up the blocks in the column. The numbers must be carefully placed on the horizontal axis - to be written beneath the *middle* of each column for a column graph, but at the line junctions if a linear graph is being created. The *transition from dealing with block graphs to linear graphs* is dealt with in a number of ways:

(a) Recording data in a data table in which each cell is identified by two labels, related to its place in a column and in a row.

- (b) (i) Games on a square grid such as Battleships, Hunt the Animals, Maps.
  - (ii) Games using the intersection point of lines, such as Cat and Mouse, Treasure Hunt, Island Positions, Hidden Shapes, Give It a Name, Noughts and Crosses.
- (c) Plan of classroom: 'Where am I standing?'

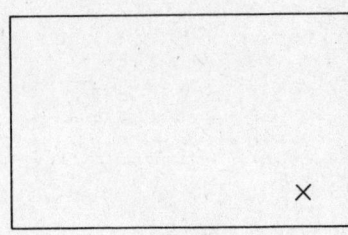

- (d) Very thin columns with a 'hat' on top, leading to no column, just the 'hat'.

By this time it is necessary for the child to be able to record the name of a point in space by its co-ordinates, the two numbers which state first how many units the point is along in a horizontal direction and second how many up in a vertical direction. The concept of *ordered pairs* naming a particular point in space is an important one. Various methods are suggested for reinforcing this concept, e.g. using geo-boards, codes, squared paper games, map making.

It is now thought by most schemes to be necessary to provide the child with plenty of opportunity to *interpret* and *interpolate linear graphs* and, in particular, to make a study of when the points in space, which are illustrating the given data, should be *linked up or not linked*, i.e. the distinction between *discrete and continuous quantities*.

Considerable time is now spent on illustrating data for which the relationship between the two variables is *linear and constant*, leading to a straight line graph, or *linear and variable*, leading to a curved graph. Some mention is made at this point of the *slope* of these graphs but fortunately no scheme (as yet) has advocated the introduction of the calculus at this age!

Some examples of graph projects demonstrating a linear relationship (with clear distinction between discrete and continuous situations) are:

- (a) constant product, constant sum;
- (b) square numbers, triangular numbers;
- (c) multiplication tables;

(d) circumference and diameter;
(e) length, area of square;
(f) ready reckoner;
(g) stretch in rubber when differing weights are attached;
(h) weight of water reaching different levels in a measuring cylinder;
(i) temperature/time graph.

The *scatter diagram* is recommended for inclusion by Fletcher, as is the *pie chart*. Certainly both of these types of pictorial representation are used frequently by the mass media and so some understanding by children is necessary. The scatter diagram might help to give a simple notion of correlation although this would not be pursued at this stage. The pie chart is attractive and readily understood, although real accuracy is difficult to attain either in constructing or interpreting the graph.

Histograms are used as a means of graphing grouped data. The word 'histogram' is *not* a name given casually to a column graph but is essentially a graph in which the data are grouped. This means it is not that the *height* of each column gives the frequency (unless all the columns are the same width as each other, and in this case their heights will be strictly proportional to the frequencies), but that the *area* of the column gives the frequency.

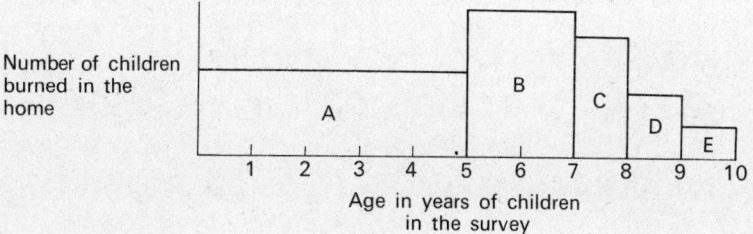

Area A represents 15 children from 0 to 5 years
Area B represents 10 children from 5 to 7 years
Area C represents 4 children from 7 to 8 years
Area D represents 2 children from 8 to 9 years
Area E represents 1 child from 9 to 10 years

Undue emphasis on *equalities* might prevent the child from realising that a frequent occurrence which may need to be graphed is that of an *inequality*.

e.g.

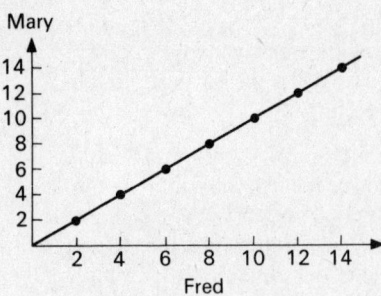

(a) Mary is the same age as Fred

Equality (any point *on* the line gives a true reading)

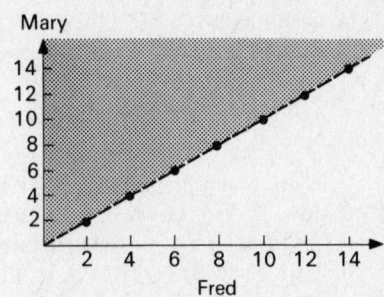

(b) Mary is older than Fred

Inequality (any point *above* the line gives a true reading)

This study of inequalities can lead to an interesting extension of graphical work which is the subject of *linear programming*. However, although some linear programming can be made to be sufficiently simple for primary school children to understand, I think that this is a subject which will be dealt with more satisfactorily by older pupils.

REFERENCE

1 *Metric Units in Primary Schools* (The Royal Society, 1970).

Chapter 3

The Changing Scene

Mathematics is a subject which has the advantage of being simultaneously organic and dynamic. In some respects it is true to say that there is nothing (mathematically) new under the sun and that a study of the history of mathematics only reveals the way in which man has investigated what already exists and has devised means of expressing his 'discoveries', adapting them for his purposes and using them to probe further. In other respects we can look on this as a process which is continually revealing new facts, new demands and new possibilities, and carving out new pathways.

The two ways of looking at a circle provide an analogy – one person will describe it as an infinite number of separate points which are equidistant from a given fixed point – another will describe it as the path of a point which moves such that it remains a fixed distance from a given fixed point. Either way of looking at it gives us the same product, a circle!

3.1 RECENT DEVELOPMENTS

The developments in the teaching of mathematics over the past few years can be described as a success story in many respects, not the least of which is that both the organic and dynamic views of mathematics are now generally acknowledged to be important and that there is a genuine attempt in most schools to take on board new ideas and approaches as considered appropriate and to throw overboard (albeit reluctantly) redundant ones. At the same time it is also recognised that there is a central core of knowledge which must survive (and is surviving) all attempts to jettison it.

Some teachers like the exhilaration of 'shooting the rapids' and tackling the unknown; others navigate among the rocks with extreme caution, planning carefully each section of the route; others find alternative calm waters which are familiar and safe. Probably the same destination is reached by successful navigators on each of these routes, the travellers having learned from the type of experience they have had.

The important thing is that each of these methods has merit to a greater or lesser extent. Just as it would be totally foolish for the non-expert canoeist to attempt to lead others into shooting the rapids, so it would be sad if the person able to shoot the rapids prevents others from sharing his experiences. The teacher must be given the task of choosing his own method of negotiating the journey while at the same time being given training, resources and encouragement to examine other methods.

3.2 INNOVATION

It becomes apparent, when one views the past few years in schools, that resistance to innovation in the teaching of mathematics is weakening. In my discussions with a considerable number of teachers there has been a recurring theme:

'A few years ago in this school the mathematics teaching was very formal. In fact it was more like arithmetic – each child had a textbook and was more or less at the same stage as most of the others in the class. The results were reasonably good – at least we were able to assure ourselves that the children knew their tables and knew how to carry out the basic operations. Teachers now prefer to have a few copies of several books for groups of children to use as they are ready for them.'

The head teacher of a large school for infants described how she changed the way of working. Her classes used to concentrate on computation of a more formal kind, interspersed with occasional lessons based on L. G. W. Sealey's series *Let's Think About Mathematics*. The classes at the top of her school liked this way of working but it was found that 'children knew what they had been taught but could not apply this to a new situation – they could not cope with a twist'. For this reason she decided, after discussions with her staff, to break away to a certain extent from the formal class teaching methods and to introduce an element of individual and group work. Concerned about a possible lack of direction and progression in this way of working and concerned also about the teachers' lack of confidence in planning such work, she decided to introduce Fletcher Mathematics as a pilot scheme into one of her two reception classes. The young teacher's verdict after working the system for some months was 'this is giving me the progression which I never knew existed'. On the strength of this, the second reception class was introduced also to the Fletcher series. When these two classes moved up, the two teachers to whom they went agreed to continue with the system even though their only encounter with the Fletcher work was in watching the progress of their colleagues with the reception classes. After twelve months of variable success the children moved on again and now two more staff have been willingly involved. Meanwhile the first young teacher is working on Fletcher Mathematics for a third time with this year's reception class and has gained in confidence sufficiently to have developed her own extensions of the work. The head teacher of this school is well satisfied with the enthusiasm of both the children and staff and it is clear that the slow change has prevented any panic or confusion.

Needless to say, innovation in other situations came in other ways. The new head teacher of a two-teacher village school was alarmed at the formality of previous work done in the school. Wanting to revitalise the teaching of mathematics in the school, but not being mathematicians, she and her colleague decided to base all their teaching on Fletcher. They made this choice because, feeling somewhat insecure about less formal approaches to mathematics, they liked the 'back-up' of the teachers' resource books. Both of these teachers are finding now that, with each new set of children 'new ideas develop, [the teachers] are more selective, they have confidence to adapt the work to individual needs and they now use it as a skeleton – a lot of practical work fits on to the bones'.

3.2.1 *Reasons for change*

When one inquires from teachers why they think it has been necessary to make changes in the style of teaching and the content of mathematics, a diversity of opinions is expressed.

(i) Some attribute the change to influences far beyond the school. In fact, even the Russian Sputnik of 1957 has been quoted as a catalyst which prompted some mathematicians in the Western world to review their own progress! Much that is new has developed within mathematics and this has caused those responsible for the content of school mathematics to take a hard look at their tradition-bound syllabuses. A number of pilot schemes have been introduced into schools to try out new ways of learning and teaching mathematics.

(ii) Others attribute the change to findings of research into how children learn. Dienes[1] describes three development stages, from a 'groping stage' in which the child is exploring, through to a period of increasing awareness which culminates in a 'moment of insight', which is followed by a need to analyse and practise the new insight. The five developmental stages of Piaget[2] need no elaboration here. His findings can readily be studied firsthand, secondhand and even thirdhand. No primary-school teacher can afford to disregard the particular emphasis he puts on the 'intuitive period' and on the 'concrete operational period'. Bruner[3] seeks 'an approach to learning that allows the child not only to learn the material that is presented in a school setting, but to learn it in such a way that he can use the information in problem solving'. He breaks the problem down into six sub-problems:

1. The attitude problem: how to get a child to extrapolate from what is already known, and to interpolate unconnected material.
2. The compatibility problem: how to get a child to fit new knowledge into what he already knows.
3. The activating problem: how to cause a child to be activated by experiencing his own success at solving a problem.
4. The problem of giving the child practice such that he has 'the tool kit for thinking'. Tools of the mind need to be used and practised to be effective in use.
5. 'The self-loop problem': how to get a child to translate for himself and retain in his own mind new experiences which he understands but cannot at first express.
6. The handling information problem: how to get a child to handle information such that it is stored, organised and recalled in an appropriate way for problem solving.

R. R. Skemp[4] emphasises a 'three-part theory for learning mathematics'. The first part recognises the dependence of mathematical thinking on reflective intelligence ('the ability of mind to turn inwards on itself') and the consequent problem of whether this reflective use of intelligence can be learned or aided by teaching. The second part emphasises the need to distinguish between the mere learning of facts and the acquisition of concepts, the latter demanding an arranged group of experiences to enable the new concept to be built upon previously acquired concepts. Thirdly, Skemp stresses the importance of schematic learning in mathematics, 'the kind of learning which makes use of, and builds more knowledge on to, an organised structure of knowledge'.

Many other psychologists, educationalists and mathematicians have been, and are, turning their attention to this vital area of research. I have found the following works of particular interest in this respect:

C. Stern, *Children Discover Arithmetic* (Harrap, 1953).
J. Piaget, *A Child's Conception of Number*, translated from the French by C. Gattegno and F. M. Hodgson (Routledge & Kegan Paul, 1952).
J. Piaget and B. Imhelder, *A Child's Conception of Space* (Routledge & Kegan Paul, 1952).
J. S. Bruner, *The Relevance of Education* (Allen & Unwin, 1972).
J. B. Biggs, *Anxiety, Motivation and Primary School Mathematics* (NFER, 1962).

One particular effect of this recent research has been the creation of various types of *structural apparatus* to provide the concrete experience needed before mathematical abstractions can be made. Some teachers' views on various types of structural apparatus, together with the creator's claims for the apparatus, are discussed on pages 111-14.

(iii) Yet other people lay the 'blame' for change on *the state of the subject* within the curriculum in relation to its relevance to the child (or lack of it), its popularity among children and teachers, and its suitability (or lack of it) as a component of a course training young people to fit into society. It was felt, not only that the content of the mathematics syllabuses contained a vast amount of outdated irrelevant matter, but also that insufficient attention was paid to the present interests and needs of the child. This was particularly true of the teacher of the child in the junior and lower secondary school where it was the practice to stress the future uses to which mathematics might be put (income tax, hire purchase and the like), neglecting the fact that to a child these were of little relevance and even less interest. On the other hand the infant school teacher has not, in general, been guilty of such insensitivity. Perhaps this is why infant teachers lead the way in showing how to help children to enjoy mathematics.

It is necessary to be realistic when considering the popularity, or otherwise, of mathematics. We cannot expect, or even desire, that everyone should think in a mathematical way, enjoy mathematical tasks or have a feel for mathematical precision, but at least we can attempt to create positive thinking about the mathematics done in the classroom rather than creating the negative attitude which comes from dislike of the methods and content. It is gratifying to read a report[5] made at the European Seminar on Mathematics Education in April 1974, that 'There has been a change in that a high proportion of pupils are arriving from primary school actually liking mathematics as an activity'. In the first part of this book I outlined a number of statements made about mathematics by teachers. Many of these statements incorporated the speaker's *aim in teaching mathematics*. Perhaps it is because of the fact that many teachers are now attempting to define their aims (although very few actually have them written down - to quote one teacher, 'Our aims are not voiced, but are quite clear. We aim to get children involved - because every involvement is gain') that changes have occurred in the style and content of mathematics teaching. For instance, it does not further the aim 'to help a child to develop strategies for solving day-to-day problems' to be demanding extensive practice at sums like this:

$$\frac{(4\frac{1}{5} \times 12\frac{3}{7}) + 5\frac{1}{3}}{2\frac{1}{2}}$$

or this:

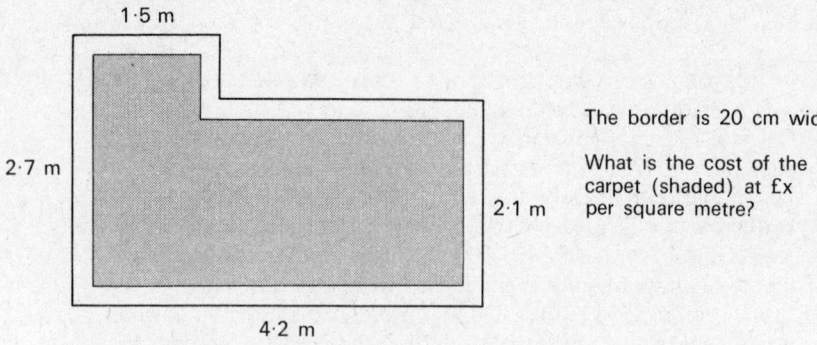

Neither can one, with an easy conscience, present the child with secondhand techniques and artificial data when one's aim is to help a child to discover for himself the relationships which exist and which determine the order and pattern of the universe. (iv) Yet another cause to which change has been attributed is *the changing relationship between the teacher and the learner.* While it has not yet become a partnership in the real sense of the word, it has become a relationship where active participation is necessary from both and where it is desirable that the child, as well as the teacher, bears some of the responsibility for initiating work. This is an outcome, not only of the research into the factors which motivate learning, but also of the belief that the child is entitled to know why he is being taught certain things. In particular, he is entitled to know the structure which lies behind the mathematics he does. To quote David Wheeler:[6] 'The teacher's task is to find ways of giving children access to their own powers. If one were setting out to look for an answer to the question "Why innovate?" it would be difficult to find a better starting place than that.'

3.2.2 *The fact of change*

The examples quoted earlier in this section of the way in which innovations were made within two schools are typical of many others, although there is considerable variation between and within schools in the methods used to create a framework on which to order the content of the mathematics done. The point worthy of note at this stage is that change in the teaching of mathematics is now a fact, not a desire of some, or a fear of many, or even a hobby-horse of a few enthusiasts. The good thing about this is that we no longer need to spend time and energy convincing people that change is necessary but instead we can plan together ways of accommodating this changed image of the subject.

3.3 THE ATTRIBUTES OF CHANGE

What are the attributes of this changed image of school mathematics? It would be unrealistic to describe these as if they are already apparent throughout our primary schools but nevertheless it is possible to identify certain characteristics related to the child, to the range of work, to resources and to the structure within the class, school and neighbourhood.

3.3.1 *The child*

The child is encouraged to think for himself, rather than being automatically taught a technique; he is given a chance to create solutions for himself. He is given opportunities to initiate work or, alternatively, to develop an idea once a starting point has been suggested, rather than fulfilling the role of follower, led by the teacher. He is helped to realise that some problems have one solution, some have many solutions and some have no solutions at all. He is encouraged to find and use available appropriate resources for the tasks he is doing and these may include his teacher, his fellow pupils, other people within the school, parents and other local people. Work occasionally takes the form of classwork, sometimes group work and sometimes individual work and whichever of these is being used it is hoped that the child is making a contribution which is causing him to extend his knowledge. Silence is still required for some individual work, but also the great value of discussion between teacher and child and within groups is recognised.

3.3.2 *The range of work*

The range of work covered is broad. It includes the traditional core aspects of number, measurement and shapes, but arising from these are topics which aim to develop a deeper understanding of the structure of number (e.g. number bases, remainder arithmetic, operation on sets), the concept of measurement and the relationships between measures, and the mathematics which help to explain how shapes arise and become useful. Underlying all the processes and activities is the hope that the child will become more observant, more thoughtful, more knowledgeable and more resourceful in matters which need these qualities.

There is the danger, of course, that by this approach we shall produce children who are 'jacks of all trades but masters of none'. This may well be so unless we have teachers who are not only numerically competent but also mathematically resourceful and sufficiently confident to motivate mathematical thinking. The children undoubtedly are influenced by the enthusiasm and involvement of the teacher and, of course, the major hold-up in schools at present is either lack of at least one person in each school who will play the major part in co-ordinating and inspiring the mathematics in the school, or the lack of a well-thought-out, long-term plan for the whole course.

3.3.3 Resources
Another of the attributes of change is the increase in the resources available and in the readjusted emphases on the use of these in many schools.

Textbooks. The class textbook is giving way either to a small number of varied textbooks in the classroom, or to a selection of individual copies of books available for child and/or teacher reference, supplemented by home-produced workcards. From their primary mathematics project[7] findings the Schools' Council comment: 'over three quarters of the children [in the forty schools taking part] use workcards that have been made in the school – the total number of teacher-hours that goes into making these must be enormous'.

Infant classes which do base their work on a text book such as the Fletcher Mathematics series or the Sealey Beginning Mathematics series, tend to favour individual step-by-step progress through the book as planned, with the teacher suggesting which sections can be shortened.

On the other hand many teachers of junior classes who like the backing of a textbook find that no textbook can expect to cater for the varying interests and abilities of every child and that a considerable amount of careful selection of sections from the textbook must be made.

Specific recurring comments about the Fletcher (*Level II*) series are that while the teachers' resource books are extremely useful, not only as a guide to progression but also as a generator of ideas, the individual children's books are not so acceptable. This is partly because the vocabulary used is somewhat specialist. Most children who have worked progressively through the series are familiar with it and can cope but it is difficult for the child who has not worked through in this way. It may be argued, of course, that it is not intended that people should get on the Fletcher train partway through the journey, but it seems to me that there is a serious limitation and lack of flexibility in any system which prevents selection of potentially useful parts. (Possibly this is one reason for the apparent lack of use of the Cuisenaire rods, in that while they are of great value to the child who has progressed through the play stage, to the colour stage, to the number stage, to the arithmetical operations stage, they are not found to be so much use to the child who has missed out on the early stages at the appropriate time.) As with almost all textbooks, the Fletcher books are considered by some to be too wordy and, faced with a page of writing and explanation, children tend to get fed up with it, take the line of least resistance and go to ask the teacher what they are supposed to do. Tedium also sets in where, with the best of intentions, children are asked to find several different ways of doing a sum. The type of mentality which thrives on this multiple approach is an adult one, and one in which work is done for work's sake. To a child the immediate goal is to get to a destination and when that is achieved it seems pointless to him to have to go over the journey again to get to just the same place. This is not to suggest, of course, that in the interests of establishing understanding, a group of children together with the teacher should not explore several ways of tackling a question.

e.g. In how many ways can *we* find what twenty-three eighteens are?

```
        18+        23+       18X      18X
        18+        23+        23       23
        18         23        360       54
    (23 times!) (18 times!)   54      360
                              ---      ---
                              414      414

        23X        23X
         18         18       (18 × 20) + (18 × 3) =
        ---        ---
        230        184
        184        230       (23 × 10) + (23 × 8) =
        ---        ---
        414        414       23 × 6 × 3 =
```

	10	10	3
10	100	100	30
8	80	80	24

414 squares

The discussion which would arise from such a group exercise would be of great benefit to many children, but even so it must be appreciated that some children would prefer to take a single-line approach. It is *not* essential to do several methods to have an understanding of a concept. What is necessary, of course, is that the one method, once established, needs to be reinforced by practice and by use.

While on the subject of tedium, one young and enthusiastic mathematics teacher to whom I spoke gets around this by sharing out

110 *Mathematics: Friend or Foe?*

the tasks suggested in a book like Fletcher and then pooling results, and also by getting the first child who arrives at a task to create the necessary chart or table on a large scale and then later-comers use the same charts. She maintains that this does not reduce understanding and considerably increases motivation and progress.

Many other criticisms are levelled at the textbooks commonly in use. These criticisms range from accusations that the scheme appears to lack coherence when the child is caused to move rapidly from one topic to another, and insufficient care has been taken by the writer to ensure that all the necessary concepts have been introduced for the topic in hand, to the accusation that primary schools textbooks are allowing children 'to take the cream off the cake too early'–or, using another metaphor, 'to pick the plums off the tree before they are ripe'. The net result of this is that children arrive at the secondary stage thinking that they have completely 'done' such subjects as symmetry and graphs and geometry (and even calculus!). While it is stimulating and exciting for the primary-school teacher to give the child opportunities to glimpse, and to get some acquaintance with, such topics, it is imperative that in her enthusiasm the teacher does not cause the child to meddle with a subject which would be far more satisfying to him if it were taken up in two or three year's time. To strike the balance between what to include and what to leave out in primary school mathematics is something which only the teacher and the members of her class can do. It will vary from teacher to teacher, from year to year, from child to child, from topic to topic. The important thing here is that there must be close liaison between teachers and between schools and careful recording of work done and progress made, so that the best time to pick the subject up again is not lost or overlooked.

A further criticism of many textbooks is that frequently tasks for the child to do are arranged such that little or no thought, together with some guesswork, are all that is necessary to get the right answer. 'Good', you may say, 'let the child appreciate the pattern–isn't this what mathematics is all about?' The danger here is that it may appear that the pattern is being appreciated (this is the stated aim of the exercise) but in reality merely a series of consecutive numbers is being filled in and the child is only needing to show his skill at counting!

e.g.

$$\Box + 15 = 15$$
$$\Box + 14 = 15$$
$$\Box + 13 = 15$$
$$+ = 15$$
$$\cdot$$
$$\cdot$$
$$\cdot$$

A further criticism of the Fletcher series is that the programme of work is out of step with most other systems (see Part 2 where the progressions have been analysed) and this provides dangers of omission and duplication where transfer is necessary from one system to another, either between schools or between classes within one school.

Structural apparatus. There is a great range of structural apparatus available. The Stern, Unifix, Cuisenaire and Colour-factor number apparatus has been supplemented by the Sealey Infants' Mathematics Set, the Dienes Multibase Arithmetic Blocks and by Centicube apparatus. Other apparatus which can be found in many schools is the Dienes Logic Blocks, the Multiboard and various types of calculators ranging from an abacus to a mini-computer! The basic aim of all of these types of apparatus is to provide the concrete experiences which underpin abstract operations. Dr Catherine Stern[8] maintains that the experimental opportunities afforded by structural apparatus are valuable and that children have a right to learn by trial and error. Some kinds of structural apparatus are self-corrective and so the child can teach himself facts which are true. When selecting which structural apparatus to use for number work, the teacher has to make a decision about the way in which he wishes his children to do the four basic operations. The various makes of structural apparatus available for number work can be classified on a continuum ranging from pure counting apparatus (e.g. the Sealey Infants Mathematics Set) at one extreme to apparatus depending solely on a length concept (e.g. Cuisenaire and Colour-factor) at the other (see Figure 3.1). For instance, the former type allows the child to count on in ones in the process of addition and to remove separate ones in the process of subtraction, while the latter stresses the structure (measured by length) of a number and hence addition is done by comparing the total length of a 'train' consisting of two or more rods with a train made up from a number of tens rods and *one* other rod.

It is interesting to note that this apparatus determines that for subtraction the method of complementary addition, as opposed to equal addition method and decomposition method, is used.

In between these two extreme types of apparatus there are many other varieties of apparatus which help to determine the methods used for the four basic operations. Perhaps their positions on the continuum might be as shown in Figure 3.2.

Dr Stern, for instance, has devised apparatus which consists of solid rods which are *grooved.* So while encouraging the child to get first the concept of the structure (in length) of a number, she retains also the opportunity of breaking the number down into ones. Hence the child *can* perform operations by counting (Figure 3.3). In her book *Children*

112 *Mathematics: Friend or Foe?*

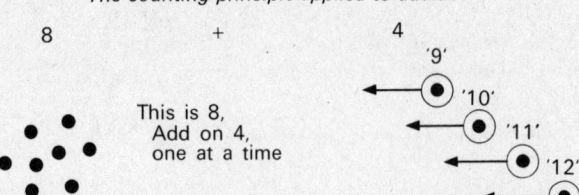

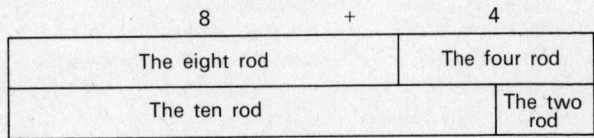

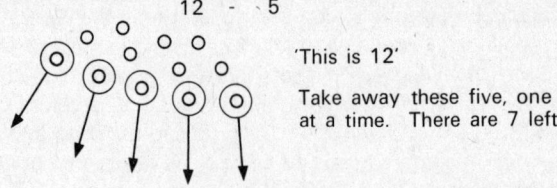

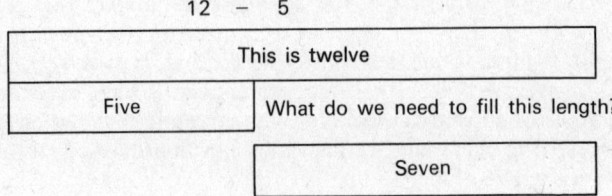

Figure 3.1

The Changing Scene 113

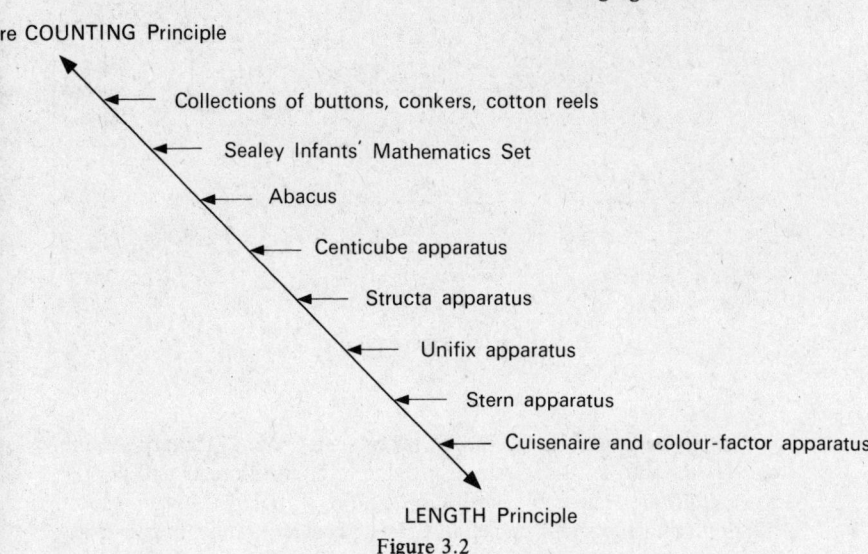

Figure 3.2

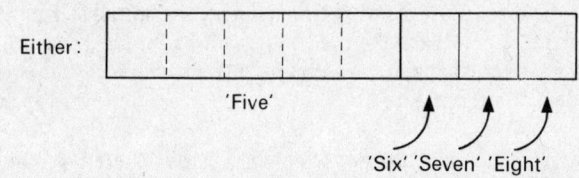

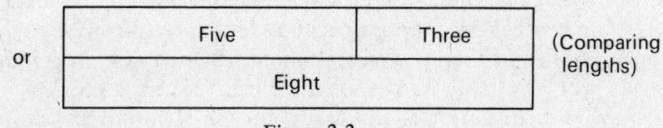

Figure 3.3

Discover Arithmetic Dr Stern claims that 'the child who has learned to think structurally has a different approach to problems from that of a child who has learned to count with counters' and that this is an important principle because mathematics has been evolved to solve problems. She stresses that the size of a number should be learned by a measuring concept first, and then counting afterwards.

Unifix apparatus attempts (successfully, in the opinion of very many teachers who use it) to get the best of both worlds. The plastic cubes can be used separately for counting or joined together to give the structure of a number. So 5, for example, can be represented in various ways (Figure 3.4).

114   *Mathematics: Friend or Foe?*

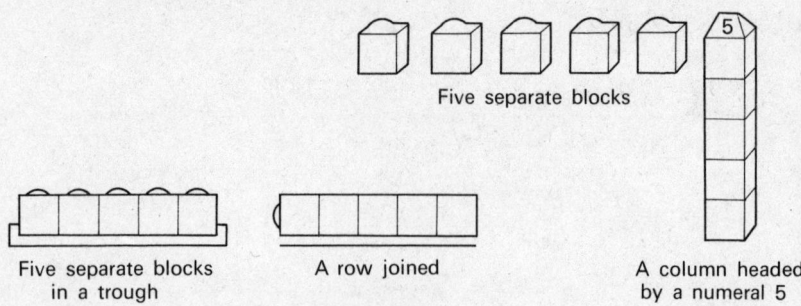

Figure 3.4

Centicube apparatus is finding its way into many primary school classrooms and, on the whole, is proving to be both useful and popular with children. The major criticism is the initial difficulty in manipulating the small cubes, but, as with ever-popular Lego, young children seem to cope with the apparatus better than most adults seem to expect them to! The advantages of their precise dimensions and weight outweigh the disadvantages of their smallness and initial stiffness. The inventors of the apparatus make many suggestions for their use in connection with number, shape, measurement and sets work, and it is unnecessary to describe them here.

Help for the teacher. Another resource which can be considered as one of the attributes of the changing face of school mathematics is that of external help for the teacher. This comes in many forms – from self-help meetings with colleagues within the school or neighbourhood, or short courses run in teachers' centres or schools, or longer in-service courses in universities and colleges of education, or BBC and ITV programmes together with related literature, or nationwide schemes such as the School Mathematics project and Maths for the Majority project, or associations such as the Association of the Teachers of Mathematics and the Mathematical Association, or bodies engaged in educational research such as the Schools Council.

With all these influences it is no wonder that the face of the teaching of mathematics constantly changes but, at the same time, it is of considerable wonder that there are still schools where any encroachment on past practices is resisted. This must point to the fact that, in spite of the list above, there is still insufficient encouragement and opportunity for teachers to attend suitable mathematics in-service courses – 'courses that really help, not ones at which teachers play at being children'.[9]

Equipment. The equipment now found in many classrooms and being used by children (in addition to the types of structural apparatus already discussed) would include many additional items compared with only a few years ago. The current most popular birthday present for the top juniors in one of the local schools is a pocket calculator and anxious parents are seeking advice on whether it is a good thing that their children should be able to bypass calculations in this way! A boy of 11, not content with mastering the basic uses of an ordinary slide rule, made for himself a simple slide rule for operating with base two numbers. *Games* of all descriptions are recognised as respectable pieces of mathematical equipment and encouragement is given to children to play fours or dice games or Mastermind or Hexapawn or with the Tower of Hanoi. Dienes Logic Blocks, for which many activities (in addition to those described by Professor Dienes and others in pamphlets issued with the apparatus and in books such as *Learning Logic and Logical Games,*)[10] have already been mentioned and are an invaluable aid to the teacher whose aims, in mathematics sessions, include the development of logical thinking in children.

3.4 THE STRUCTURE WITHIN THE CLASS, SCHOOL AND NEIGHBOURHOOD

The structure within which change is occurring is itself an attribute of change, not only in the teaching of mathematics, but in all aspects of teaching. The inter-relationships and interdependence of the head, the teachers, the neighbourhood, the child, the parents, the authority, the local employers, the planners, are recognised as factors which must influence and help to determine what goes on in the school. While, in theory, each head teacher is autonomous in his own school, in practice he is constrained by public opinion, by accepted modes of behaviour in society and by common practices (for instance it is now considered a normal practice that the equipment needed by the children is provided by the school). All these features affect the type of mathematics being taught. There is frequently a demand from parents, who in past years have been content to accept that school knows best, that their children become proficient at doing the mathematics needed for 'everyday life', and for passing qualifying examinations. Balanced against this, there are the exhortations of educational researchers who seek continually to enrich the learning process and put less stress on the utilitarian and vocational aspect of mathematics. In the middle of these is the practical teacher who seeks to guide a class of children through a meaningful, motivating, relevant learning experience. How can the content and styles of mathematics teaching remain stagnant under these pressures? The answer is that they cannot do so and they do not do so.

3.5 THE VEHICLES OF CHANGE

Having thought about the attributes of change perhaps we should consider what are the vehicles for change in order that we can assess, in some measure, the contribution each one can make towards more effective teaching of mathematics.

3.5.1 *The individual*

Without doubt the major vehicle is the individual enthusiast in the classroom. A headmaster once told me that whatever standard a teacher sets for himself, the children will reach just below that standard. This is a forthright statement but it has a lot of truth in it; and on the subject of the degree of enthusiasm engendered it could well apply. Children are sensitive to the teacher's attitude to what he is teaching and, particularly in mathematics, his sense of enjoyment and enthusiasm and spirit of inquiry and interest in the outcome will generate the mood and attitude of the children to it. Of course there will be exceptions to this at both ends of the scale - children whose enthusiasm goes beyond that of the teacher and who initiate lines of inquiry for themselves, and also children who steadfastly resist being 'switched on' in mathematics and whose energies are directed elsewhere. It is sometimes easy to lose sight of the fact that children have their own opinions about how they like to learn mathematics. In the Macpherson, Harris, Leaven article[11] entitled 'How I Like to Learn Mathematics' the comments of several children are recorded:

'No future reference to Piethagorse - as he spoilt maths for me.'
'I like to learn mathematics bit by bit or I get confused. I like it to be explained thoroughly before I start to work.'
'To start with when I learn a new topic I like to be told about what uses it has, how it is connected with other subjects and generally anything of interest about it.'
'Fluently, gabbley, waffley and subscriptibly. I prefer the fluently method.'
'Mathematics is a peculiar subject it can be hard, it can be easy, it can be exciting or dull. It depends a great deal on the teacher's approach and influence and the way he or she teaches it to make the child see the exciting or dull side.'

So if we are to improve the teaching of mathematics we must seek first to create enthusiasm among people who are going to teach. In general people like doing what they can do. Does this then point to the fact that there are too many people teaching mathematics in the primary school who are not able to generate enthusiasm because they have not sufficient confidence in their own mathematical ability? It is encouraging that such bodies as the Committee for Education of the CNAA[12] are insisting that all prospective teachers must demonstrate adequate numerical competence:

'The Council makes a general requirement that any degree award shall be made only in respect of a course which includes some measure of general education and is particularly concerned about standards of literacy and numeracy. This is particularly important in any course designed for the education and training of teachers and colleges are required by the Council to make explicit guarantees of attainment in these respects in any course submitted.'

So is it true that only people with a minimum qualification in mathematics, e.g. O level, should be admitted on to training courses for primary school teachers?

For the teacher who is vitally interested in the teaching of mathematics there are many opportunities to become more proficient and knowledgeable and to share ideas and experiences with others.

3.5.2 *Physical resources*

Other vehicles for change are, of course, the physical resources available for the classroom, apart from the textbooks, structural apparatus and games already mentioned. Outside resources must include such valuable sources as radio and television, schools' loan services, services offered by such concerns as the Bank Education Service, the Schools Council projects. No teacher need feel on his own in this. Unfortunately many teachers are reticent about enlisting the help of such resources. Why? The reasons given to me by teachers include the following:

(a) The teacher is concerned that by sparing time fitting in other things into the syllabus, he cannot cover his normal syllabus.
(b) It is easier to go on teaching in the routine way, rather than attempting something new.
(c) Information is not readily available for the teacher to know how to contact the resources.
(d) There are insufficient opportunities for the teacher to become familiar with audio-visual aids.
(e) Courses for teachers are held in inconvenient places and at inconvenient times. More courses should be held in school, with perhaps two or three schools in the locality co-operating, in school time.
(f) The teacher has no free time during the working day (when planning can be done while the appropriate people can be

contacted) and so, by default, the arrangements are left for 'another time'.

(g) The teacher is apprehensive about using a teaching resource which may require specialist help to set up and to maintain. In this respect there should be one person in every school (not necessarily non-teaching) who could be available and competent to deal with such contingencies.

(h) The practical details of booking the apparatus, freeing the appropriate space, carrying the cumbersome equipment and moving the children take up too much time and disrupt the smooth running of the class. The time can better be spent in working peaceably, happily and traditionally in the classroom.

(i) The children like to do mathematics in which they know they are secure, preferring to do a page of sums rather than tackling unfamiliar activities.

(j) The teacher has tried various resources but found them to be unsatisfactory in a number of ways. For instance, criticisms levelled at schools' television broadcasts include:

  (i) The subject matter covered in a programme is too wide and it is difficult to appreciate the coherence of the total year's programme.
  (ii) Too much is tackled at too superficial a level.
  (iii) The subjects being tackled do not fit in at the right times with the more formal mathematics that the children need to be doing.
  (iv) The pace of the programme is too slow.
  (v) The pace of the programme is too fast.
  (vi) The presentation of the programme, in attempting to speak to a broad age range, is bewildering to some, patronising to others and just right for a few.
  (vii) Too often the full advantages of the medium are not used and a programme is transmitted which could just as easily have been given by the teacher, with the notes. The TV medium should take the opportunity of bringing into the classroom experiences which otherwise would be excluded. (In this respect I feel that full marks should go, for example, to a programme such as the one on 'Addresses' in the ITV mathematics series for 7+-year-olds, entitled *Figure It Out*. This programme combines an effective teaching of the concept of graph co-ordinates with relevant views of specialist maps, aerial views, station luggage containers and a computer–all of which would have been difficult to produce in the classroom.)
  (viii) It is not convenient to be restricted to watching the

programme at the given time. Other events in school often preclude the TV programme. The use of video tape recordings may remove this difficulty.

3.5.3 *Organisational structure of the school*

One of the major vehicles which has caused change in the style of the teaching of mathematics is the organisational structure of the school. It is now rare to see a primary school classroom in which there are rows of desks. In many, the desks have been replaced by flat-topped tables supplemented by a store in the room for the child to keep his books. The rows have been replaced by groups of tables. Children's chairs are facing in various directions and not all facing the blackboard. The teacher's table is still a focal point in the room but is not necessarily in a commanding position. The apparatus is classified in areas and available for immediate use rather than being kept out of sight in the cupboard. This geographical transformation has had inevitable consequences upon the style of teaching. While some class teaching is still retained as essential by most teachers, the value of a group of children working together and discussing an activity is not only accepted but planned. In response to a question put to a large number of teachers – 'What form of organisation do you normally prefer when teaching mathematics: mostly individual work, mostly group work or class teaching?' the almost unanimous answer was 'a combination of all three'. In classes homogeneous in age, many teachers are still attempting to keep most of their children working on a particular topic, but at varying stages within the topic. This means that the introduction to a topic is done by class teaching but thereafter children work in groups or individually, developing the topic as far as they can before the whole class moves on to a new topic.

One teacher's analysis of some approaches used in the classroom is shown in Figure 3.5.

To quote another teacher, 'Class teaching is the only way of being sure that everyone is with you'. The class teaching approach is favoured by many as the best way to start the mathematics lesson anyway, in order to set the atmosphere for the session. The discussion may be connected with the new topic or alternatively a one-off discussion on, for example, 'My favourite prime number' or powers of numbers and the 'rice on the chess board' puzzle. Another teacher remarked 'There are always some spin-offs from a class discussion and children do remember more than we think they do from them'.

Data in *What's going on in Primary Mathematics*? is given in an interesting diagram (Figure 3.6; data extracted from forty schools in England and Wales in 1974).

120 *Mathematics: Friend or Foe?*

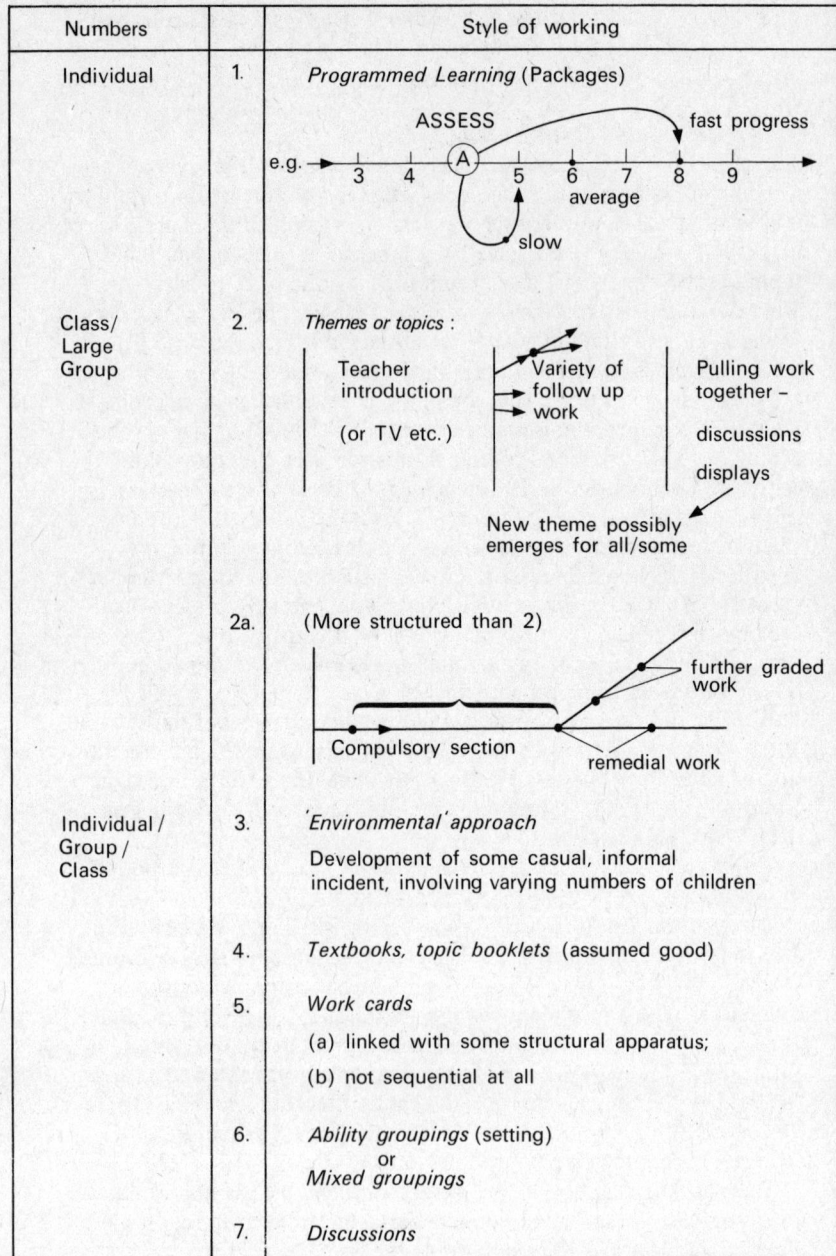

Figure 3.5
With acknowledgement to J. F. Porter.

(For instance, we might estimate from this diagram that the class work data shows that:

Approximately 4 schools submitted no data,
3 schools do no classwork,
20 schools do less than half their work as classwork,
6 schools do about half their work as classwork,
6 schools do more than half their work as classwork,
1 school does all classwork.)

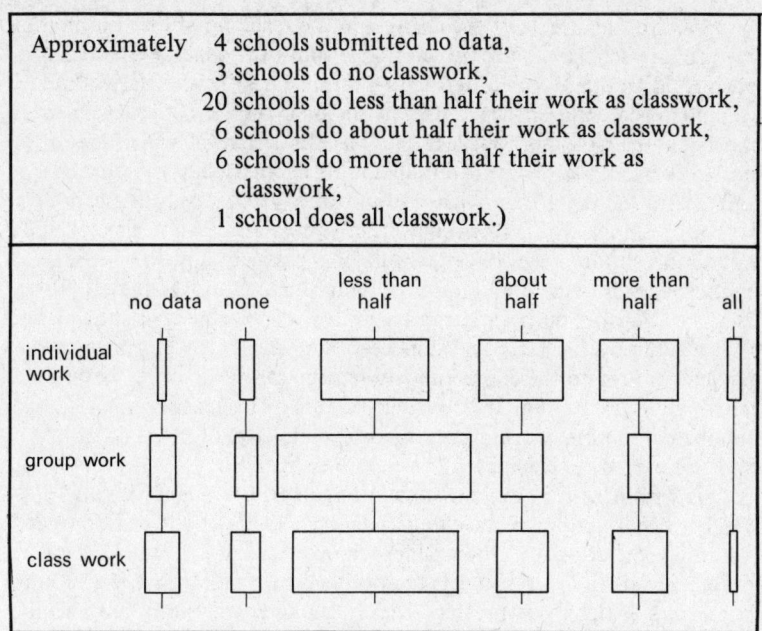

Figure 3.6 *How much individual, group and class work?*

In the smaller one-teacher or two-teacher schools, class teaching is neither desirable nor necessary and the teaching is with individuals or groups. In one particular school where an integrated day system has been established, the children plan their own schedule of work, but 'They can choose when but not what...and they become good at disciplining themselves....This approach is very demanding, but so much more rewarding. We all have to be desperately tidy so that everyone can find what they want when they want it....' Occasionally this school has a break from the individual work and the whole school concentrates on one theme, perhaps suggested by a child or a chance happening, e.g. 'If I gave you £100 what could you buy?' or 'Trees' or 'Triangles' (including a boat trip on the river to spot as many triangles as they could see!)

The approach which needs to be used in these contrasting situations is determined by the size of the school and by the philosophy of the head teacher. These in turn determine the scope of mathematics within the school.

3.6 THE EFFECTS OF CHANGE

3.6.1 *General effects*

So what are the effects of the changes in the teaching of mathematics in the primary school? In accord with one's aims, the general effects are both good and bad! The orderly, quiet atmosphere of the classroom in which (most) children were working on their own has given way in most schools to a more ragged, less peaceful situation which has probably needed far more careful planning and certainly more careful and skilful handling than with the previous system. An unsuspecting visitor to the classroom might have the impression that the classroom is noisy, the children are moving about unnecessarily and that little serious work is being done. In some circumstances, unfortunately, this may well be true, but the perceptive visitor will be able to distinguish between a working buzz and unnecessary noise, between purposeful movement around the classroom and unnecessary wandering, between worthwhile use of time and wastage of time. The teacher must aim to create a genuine working atmosphere in all parts of her classroom and this can only be achieved through the positive co-operation of all people within. When children are working in groups or individually the teacher needs 'antennae' reaching to all parts of the room and, to quote a leading educationalist, the teacher who concentrates too exclusively on one group and who neglects the others is helping to train the future occupants of the modern office – that is, the ones who work only while the boss is watching! The task of ensuring that all individuals and groups are progressing demands careful management and is sometimes achieved in one of the following ways:

1. Only one group at any one time is occupied in mathematical work of an experimental nature which may require more of the teacher's time than can normally be spent. Other children are consolidating work done previously, practising a skill or completing computation. A class textbook is sometimes used to provide this work.
2. A limited number of children are working at mathematics at any one time. Other children are occupied on activities which, at that time, require less teacher involvement.
3. Two teachers work as a team in order that one can afford to involve herself with a group of children doing mathematics while the other teacher takes responsibility for a large number of children.
4. The teacher is an opportunist who grasps occasional opportunities during the week to initiate appropriate group work.
5. Group work is initiated by a class discussion and then each group works on a similar assignment so that the teacher can monitor progress by comparing groups and so that children who succeed readily can help others.

Another general effect of the change is that mathematics does not instil as much dread in children under the age of 12 as it used to do. The increase in the popularity of mathematics with younger children is encouraging and rewarding to the primary school teacher who is saddened though by the loss of prestige and popularity which the subject suffers by so many at some time within the subsequent few years. Why is this? Perhaps the secondary school has been less susceptible to change, more tenacious of the traditional, formal approaches to mathematics and of the generalised techniques and skills which have been the content of public school curricula for some generations.

3.6.2 *Specific effects*

The teacher. There have been specific effects upon the teacher. He has lost some of his security and complacency (and boredom?) and substituted for them a degree of uncertainty and inquiry. What was organisationally easy is now far more difficult. The teacher is learning while he is teaching and is not only teaching the things he knows. He is becoming more knowledgeable and has more scope for adopting his own style of working rather than that imposed by rather restricted and inflexible methods, and content. At the same time the teacher is becoming more dependent upon others in that he cannot, as in the past, only teach a certain compact section of the syllabus in isolation, but his work depends upon the attitude, application and understanding instilled previously. He, in turn, is establishing attitudes and emphases for the future. He is also faced with the difficult task of keeping *a record of the progress* of children working in different ways, at different levels, and on different tasks. The days of the mark book with its neat, complete sets of marks for a certain piece of work are becoming a thing of the past (but are not entirely gone).

The child. What are the effects upon the child? Words and phrases which occur frequently in arguments about children's work or attitudes in mathematics range from criticism that children are 'careless', 'fed-up', overwhelmed, too superficial, lacking in real hard work, 'wasting time'–to praise that they are more resourceful, inquiring, independent, interested, knowledgeable. One wonders whether mathematics teaching can be likened (with a little poetic licence) to the nursery rhyme:

> 'When she was good she was very, very good,
> But when she was bad she was horrid!'

Some primary school teachers find themselves still being constrained in their mathematics teaching by the demands of qualifying examinations, entrance examinations or the need to get children to a certain level of competence to compare favourably on entry to secondary school with children from other schools. Consider the dilemma for instance, of the head teacher who realised that one of her most promising pupils with regard to activity mathematics failed the entrance test to the local direct grant school as she was rated as 'very much below average'. Further inquiry revealed the fact that the 9-year-old girl did not recognise a long division sum, although when the problem was expressed in words she was able to do it in her head, and she didn't do the long multiplication because there was not enough room to do it in her usual (Fletcher) way, so she thought the question must mean something else! What was at fault? The child? The teacher? The test?

Authors of textbooks. Mathematics now requires a greater proficiency in reading. Writers of textbooks for children have been battling with the difficulty of keeping parallel the levels required in both reading and mathematics. To help the teacher it is now not uncommon for authors to list the vocabulary required within a mathematics textbook and this practice has gone some way towards solving the problem. However, another problem is to strike a good balance between the quantity of words used and the amount of mathematics involved. Faced with a page of script, the child cannot be blamed for taking the line of least resistance and asking the teacher what has to be done. *I do not think that, in general, a primary school child should be required to teach himself mathematics through the written word. This is not the main purpose of the textbook.* In my opinion the purposes served by a textbook are:

(a) to provide a comprehensive coverage of mathematics appropriate to the level (in this respect the chief benefit is to the teacher);
(b) to supply an adequate set of questions to test and reinforce techniques;
(c) to suggest starting points for other investigations related to the topic;
(d) to provide a source of work for children to be able to continue to work without the teacher's involvement during those periods when the teacher is necessarily involved with another group or another task;
(e) to give the parent an opportunity to follow the work done by the child;
(f) to give the child who is eager to develop some work on his own some material to work on;

(g) to provide, where necessary, some worked examples;
(h) to reproduce for the teacher and the child those diagrams, charts and pieces of data which would take too much 'mathematics' time to devise or collect;
(i) to suggest and develop links between various branches of mathematics, e.g. number and shape, shape and measurement, measurement and number...

Many textbooks at present in use in schools fail on some of the above counts. It is difficult to perceive how, if at all, the isolated topics relate to each other and how the total programme coheres. There are insufficient reinforcement questions. Sections are too wordy and attempt to do the work of the teacher. Too much effort is put into forcing relevance into tasks with the result that some simple mathematical concept is disguised in a phoney 'real-life' situation. The plea for relevance is over-rated and is not echoed by young children. So often our attempts to force in relevance leads to a ridiculous fabrication of tasks about carpets or holidays, and at the same time perpetuates the impression that it is a waste of time to do anything else in mathematics.

Teacher training. The training of primary school teachers of mathematics has undergone considerable change over the past few years. Of all the subjects offered as part of a teacher training course it is probably true to say that mathematics is the one in which attitudes and, to a certain extent, ability tend to be polarised on entry to college more than in any other. This has a number of interesting effects. Most students recognise readily that mathematics is a core subject and that, as primary school teachers, they must equip themselves to teach it no matter what their own ability and inclinations are. So initially there is very little resistance to the compulsory part of the professional studies course. Many students find that when they are trying to decide how to present mathematics to young children they find a new or revitalised interest in it themselves and also they find that studying the structure of a system gives them a clearer overall view of what previously had appeared to be isolated techniques. However, students are highly critical of the content included in some schemes of work, textbooks or television programmes and one finds a reluctance, or fear, to introduce new untried topics. They appreciate that the wider view of mathematics has more chance of involving people within a wide range of interests.

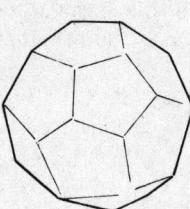 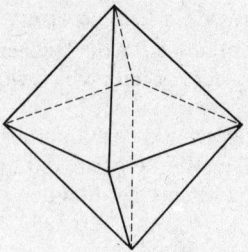

For example, the interest of the art lover can be caught by studying natural and man-made shapes, the attention of the biologist can be held by a consideration of the relationship between surface area and volume:

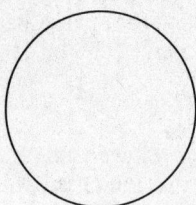 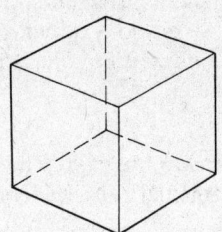

If these have the same volume as each other, which has the smaller surface area?

The 'thinker' will be given plenty of scope within the logic activities and in problem solving. The craftsman will find outlet in considering and making models and in the mathematics which lie behind them.

Convergent, orderly thinkers who like to come to a conclusion or decision can appreciate the order and pattern within mathematics. The divergent thinker has abundant scope for branching away from the central theme and pursuing his own fancies. Perhaps all of this can be summarised (in an often-used phrase) by saying that there are abundant diverse opportunities within mathematics to 'switch people on'.

Mathematical content. The effect of change on content? Sufficient has been said already about this. The undeniable fact is that the image of mathematics is changing from a subject which was tolerated (or even disliked) to a subject which is enjoyed. Speaking of the content of the Fletcher Mathematics books, for instance, one headmaster said 'So long as we have Fletcher Mathematics we shall have no children who don't like mathematics'.

Methods of analysis and recording in the classroom
As well as making the organisational changes already described (page 119), the teacher must plan carefully:

(a) how he is going to diagnose the child's needs;
(b) how he intends to assess the child's progress;
(c) how he is going to record the child's progress.

Various systems are continually being produced by various local authorities and teachers' groups.

In connection with (a) there are available a number of objective standardised diagnostic tests and the results from these can help to modify or reinforce the subjective opinion of the teacher. To assess written computation is comparatively simple and it can be done objectively but it is inevitable that the teacher's judgement, based on an intuitive norm, will provide the assessment of non-verbal and non-written work. The task of recording his results and impressions for each child is a lengthy and time-consuming one but is essential if there is to be a safeguard that in this loosely structured framework there is to be an underlying coherence and progression. A number of local authority advisors and groups of teachers have attempted to create suitable record sheets. These vary from a full report on an individual's acquisition of concepts, one by one, to an outline list of all the topics which need to be covered. An example of the first is the Worcestershire Mathematics Progress Guide, a short part of which is reproduced below:

TIME

Can the pupil:
1   Differentiate between now, and the past and the future?
2   Tell the sequence of his day (e.g. the order of his meals)?
3   Recognise special times in the day?
    ...
9   Tell time to the nearest minute?
10 Measure time in seconds for simple activities?
    ...
17 Read and use times from bus-tables and the *Radio Times*?

Or here is another example from the checklist of objectives in primary mathematical skills and concepts, entitled 'What should our children be able to do?' (produced by a group of teachers at a teachers' centre):

STAGE ONE

Name..has shown that he/she is able to do the following:

LENGTH

Use the words longer than, shorter than
wider than, narrower than
taller than, shorter than

...

Measure   using spans
using strides
...
make guesses and see if they are right

CAPACITY

Use water to compare one vessel with another
Use water to find two vessels which hold the same amount

Example 3 is from a primary school scheme:

Name..   Date of birth..................

Number Progression

1 Addition and subtraction to 20
2 Sorting into sets up to 20 (division)
3 Counting in 2s - odd and even numbers
...
14 Equivalence of fractions ($\frac{1}{2} \rightarrow \frac{1}{16}$)
...
29 Prime numbers
...
31 Averages

Some teachers prefer to use only a brief general report on each child, for example:

....School   *Mathematics Record*
Name...   Date of birth..................

Date........................ Fletcher Level(s)............ Book(s)..........
Areas of Special Difficulty:

Other comments:

Most teachers have devised some recording system of their own but this *ad hoc* 'bush telegraph' type system has led to a serious lack of communication between one teacher and the next and this has been partially responsible for the ragged impression which many people have of modern mathematics methods and for the bewilderment of many teachers about the part they need to cover in the overall learning of mathematics by the child. Many people, including teachers, are worried and confused and ask 'What *is* going on in school mathematics these days? When I was 8 I knew all my tables and I could do long multiplication and long division.' We must be careful that we shall be able to show that children leaving school now know more, think more and enjoy more. If they don't then we must be prepared to admit that our methods are still no better than they were.

Parents. The parents too have been affected by changes in the teaching of mathematics. While some are baffled and worried that their children will not reach a required standard, 'the parents who know some mathematics are happy to go along with it' (to quote one head teacher). Some young parents were themselves taught by the more recent progressive methods and in a few years' time the gap between the parents' and the child's kind of mathematics will have closed somewhat. Meanwhile schools are attempting to bridge the gap by having parents' meetings, creating occasions for the parents to work with their children. Parents now could involve themselves with the same work by watching the same television mathematics programmes as their children, (although very few in practice do so). Numeracy is the major thing that parents require the mathematics teacher to provide for their children. In this respect most of the parents can easily be persuaded that the new approaches aim to give their child understanding of a lot of related facts about, say, 12, rather than a few facts held in separate compartments.

e.g.

$1 \times 3 = 3$ $1 \times 4 = 4$
$2 \times 3 = 6$ $2 \times 4 = 8$
$3 \times 3 = 9$ $3 \times 4 = 12$
$\underline{4 \times 3 = 12}$ $\overline{4 \times 4 = 16}$
$5 \times 3 = 15$ $5 \times 4 = 20$

What a pity it is that *rote learning* of tables may prevent a child from linking up related facts like these. Not, of course, that there is not a place for rote learning in the mathematics lesson. It may well be the *only* method by which a particular child can grasp essential number bonds. If this is so, then the end justifies the means – for that one child. Things which have been thoroughly learned are stored in the memory for future use. If rote learning serves a purpose here then I'm all for it – provided that we keep it in perspective! Almost all the teachers with whom I have discussed this subject of rote learning do not reject it as a method of learning useful to some children and even enjoyed by them.

3.7 THE VERDICT SO FAR?

As I see it, then, the verdict on the new approaches of mathematics teaching of the past few years is one of guarded approval. If the subject is more popular than it was, then children and teachers will be more motivated to succeed. To be realistic one must agree that, while some children *can* generate their own work, it is still the teacher who, in most classrooms, initiates or sets the tasks to be done. Teachers are ready to be influenced but in the end they must (and will) settle into their own style of teaching. A teacher must be committed to whatever manner of teaching he adopts. It is impossible to try to act a false role with a group of young children – they sense the artificiality of the situation and are no more taken in by it then the teacher is himself. However, no matter what approach the teacher uses, one important transformation that has happened is that children are not merely thought of as 'bottles to be filled' but also as 'candles to be lit'. When given a certain amount of responsibility for their own progress and programme, children do respond, by creating a classroom in which there is not just one teacher and thirty children, but many more 'teachers' and thirty-one learners!

3.8 AFTERTHOUGHT

Story 1: A 6-year-old boy, when asked what eight fives were, went away to think and returned a little later to say that he thought the answer must be the same as four tens.

Story 2: An 8-year-old girl did not know a set method for doing long multiplication and so devised a method for herself by splitting up the multiplier and using a form of doubling-up to get the result.

Story 3: A 6-year-old girl, when told by her mother that the four members of the family were sharing a five-egg omelette, thought for a moment or two and then said that they would each have one and a quarter eggs.

Do we tend to make our subject appear more difficult than it really is? Do we shroud it in a kind of mystique? Might it be that we often confuse the issue by producing too many polished techniques and that the teacher who told me 'somehow they can learn more mathematics when they don't know any methods' was right? Could it be that, by using methods which are too structured and too bound up with the traditional method of doing things, we limit the resourcefulness and insight demonstrated by these three children?

REFERENCES

1 Z. P. Dienes and E. W. Golding, *Modern Mathematics for Young Children* (ESA, 1965).
2 J. Piaget, *A Child's Conception of Number,* translated from the French by C. Gattegno and F. M. Hodgson. (Routledge & Kegan Paul, 1952).
3 J. S. Bruner, *The Relevance of Education* (George Allen & Unwin, 1972).
4 R. R. Skemp, *The Psychology of Learning Mathematics* (Penguin, 1971).
5 Report in *Mathematical Education in Science and Technology* (April 1974).
6 D. Wheeler, article in *Trends in Education,* No. 23 (July 1971).
7 The Schools Council, *What's Going On in Primary Maths?* (The Schools Council, 1974).
8 C. Stern, *Children Discover Arithmetic* (Harrap, 1953).
9 The Schools Council, *What's Going On in Primary Maths?*
10 Z. P. Dienes, *Learning Logic and Logical Games* (ESA, 1970).
11 Association of Teachers of Mathematics, *Focus on Teaching* (Association of Teachers of Mathematics, 1973).
12 Council for National Academic Awards, document dated 17 October 1974.

Appendix

Testing

Many standardised tests have been devised for evaluating mathematical ability and progress. The teacher is only one of the many people to whom these tests are extremely important. It may be useful to remind ourselves of some of the functions of testing which will directly benefit the teacher in his quest to do the best for the child.

The functions of testing include:

(i) *Testing for classroom purposes*
(a) to diagnose learning difficulties of individual children,
(b) to help to discover the potential of each child,
(c) to identify discrepancies between the learning potential of the child and his achievement,
(d) to assess improvement,
(e) to group children within the class in whatever form is required,
(f) to guide the teacher in the planning of future work, and
(g) to identify children who need special help in given situations.

(ii) *Testing for guidance*
(a) in discussions between the teacher and the parent,
(b) in helping the child here and now,
(c) educationally and vocationally, and
(d) in helping to identify problem cases.

(iii) *Testing for administrative purposes*
(a) to form the desired groupings (unmixed, mixed...) within the school,
(b) to place new children in an appropriate group,
(c) to evaluate teaching methods and content,
(d) to provide, when required, information to outside agencies such as other schools to which the child is being transferred.

R. L. Thorndike and E. P. Hagen[1] stress the importance of the tester asking himself certain questions, before doing any testing, such as 'What information do I need that I do not have now?' 'When do I need it?' and 'How will I use it?' They make the observation: 'Defining functions and purposes [of testing] is the horse. Let us put him out in front, and the cart carrying a programme of tests will follow after.'

The confidence with which the teacher uses tests with his children will depend on such factors as the suitability of the test for the purpose, the conditions under which the test is set, the correct application of the test, the careful scrutiny of results and the interpretation of the results. If the teacher is using a recognised standardised test then it is certain that considerable research has contributed to the composition of the test. A typical programme for devising a test might be summarised by Figure A.1. It is necessary, of course, for the tester to have at least a fundamental knowledge of statistics to enable him to read the results with a measure of understanding and to interpret their significance. A useful book in this connection has been written by A. C. Crocker.[2]

Tests which measure attainment or tests which diagnose particular learning difficulties are available. Some tests are devised for individual application, while others can be given simultaneously to children in a group.

Listed below are some tests which have been found to be useful. This list is by no means exhaustive.

Tests supplied, with a manual of instruction, by the University of London Press Ltd, St Paul's House, Warwick Lane, London, EC4P 4AH
Graded Arithmetic-Mathematics Test, devised by P. E. Vernon, MA, Ph.D, decimal currency edition, July 1971. This test is diagnostic, for the age range 7 to 21, is applied individually or to a group, and takes 20 minutes.
Group Mathematics Test, devised by D. Young, BSc, 1972. Two forms of the test, A and B, are supplied, these being parallel tests to be given to a group or to an individual, within a week of each other, to a wide range of ability of children between the ages of 6 years 6 months and 8 years 6 months and to less able children up to the age of 12 years 11 months. The test takes approximately 45 minutes.
The Leicester Number Test, prepared by C. Gillham, BA, Dip. Ed., Dip. Psych., and K. A. Hesse, 1970, for the age range 7 years 1 month to 8 years 1 month.

Tests supplied, with a manual of instruction, by Oliver and Boyd Ltd, 39 Welbeck Street, London, W1
Schonell Diagnostic Arithmetic Tests (Updated). There are twelve tests, to be given to children between the ages of 7 and 14, covering all the basic processes of addition, subtraction, multiplication and division of numbers.

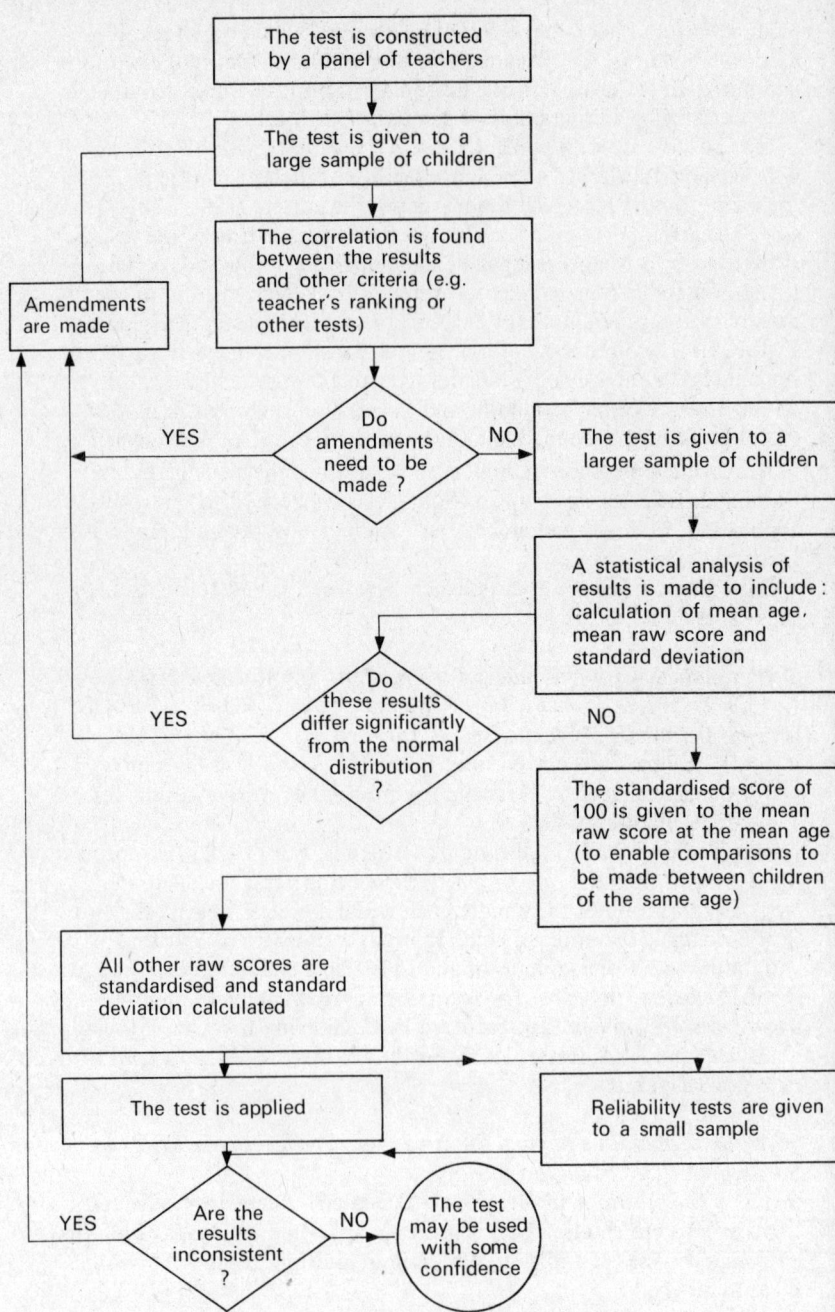

Figure A.1 *How a test may have been devised.*

Tests supplied, with a manual of instruction, by the National Foundation for Educational Research, The Mere, Upton Park, Slough, Bucks.
Basic Mathematics Tests A, B, C, D, devised in 1971/72, for first-, second-, third- and fourth-year juniors respectively.
Mathematics Attainment Tests A, B, C, D, devised since 1965, for first-, second-, third- and fourth-year juniors respectively. The first-year test is an oral one, in which the tester reads the question which applies to the diagrams on the child's printed test sheet. Within the categories A, B, C and D other tests are classified.
Number Test DEl (formerly Number Test 1), 1965, 1971, for children between the ages of 10 years 6 months and 12 years 6 months.

Tests supplied, with a manual of instruction, by Methuen and Co. Ltd, 11 New Fetter Lane, London, EC4
Primary Mathematics Diagnostic Tests, devised by J. S. Flavell and B. B. Wakelam, 1966, for the age range 10 and over.

REFERENCES

1 R. L. Thorndike and E. P. Hagen, *Measurement and Evaluation in Psychology and Education* (Wiley, 1969).
2 A. C. Crocker, *Statistics for the Teacher* (Penguin Books, 1969).

Bibliography

ATM RESEARCH GROUP, *Focus on Teaching* (ATM, 1973).
BERGAMINI, D., *Mathematics* (Life Science Library, 1970).
BIGGS, E., 'Metrication in the School Curriculum' in *Trends in Education*, No. 26 (April 1972).
BIGGS, J. B., *Anxiety, Motivation and Primary School Mathematics* (NFER, 1962).
BRUNER, J. S., *The Relevance of Education* (George Allen & Unwin, 1972).
CROCKER, A. C., *Statistics for the Teacher* (Penguin Books, 1969).
CUNDY, H. M. AND ROLLETT, A. P., *Mathematical Models*, 2nd edition (OUP, 1961).
DIENES, Z. P., *Learning Logic and Logical Games* (ESA, 1970).
DIENES, Z. P. AND GOLDING, E. W., *Modern Mathematics for Young Children* (ESA, 1965).
EDWARDS, I. E. S., *The Pyramids of Egypt* (Pelican, 1961).
ESCHER, M. C., *The Graphic Work of M. C. Escher* (MacDonald, 1967).
'European Seminar on Mathematics Education' in *Mathematical Education in Science and Technology* (April 1974).
FLETCHER, H., *Mathematics for Schools, Level I* and *Level II*, and the Teachers' Resource Books (Addison Wesley, 1971).
GARDNER, K. L., GLENN, J. A. AND RENTON, A. I. G. (eds), *Children Using Mathematics* (OUP for ATCDE, 1973).
GARDNER, M., *More Mathematical Puzzles and Diversions* (Pelican, 1966).
HOLT, M. AND DIENES, Z. P., *Let's Play Maths* (Penguin Books, 1973).
HOLT, M. J. AND MCINTOSH, A. J., *The Scope of Mathematics* (OUP, 1968).
HOOPER, R. (ed.), *The Curriculum: Context, Design and Development* (Oliver and Boyd, 1971).
KIDDERMINSTER TEACHERS' CENTRE, *What Should Our Children Be Able To Do?* (June 1971).
LAND, F. W. (ed.), *New Approaches to Mathematics Teaching* (Macmillan, 1969).
MATHEMATICAL ASSOCIATION, *Primary Mathematics, A Further Report* (Bell, 1970).
PIAGET, J., *A Child's Conception of Number* translated from the French by C. Gattegno and F. M. Hodgson (Routledge & Kegan Paul, 1952).
PIAGET, J. AND IMHELDER, B., *A Child's Conception of Space* (Routledge & Kegan Paul, 1956).
POLYA, G., *How To Solve It* (Princeton University Press, 1957).
THE ROYAL SOCIETY, *Metric Units in Primary Schools*, (The Royal Society, 1970).

THE SCHOOLS COUNCIL, *Mathematics in Primary Schools*, Curriculum Bulletin No. 1 (HMSO, 1965).
THE SCHOOLS COUNCIL, *What's Going On in Primary Maths?* (Schools Council, 1974).
SEALEY, L. G. W., *The Creative Use of Mathematics in the Junior School* (Blackwell, 1966).
SEALEY, L. G. W. AND GIBBON, V., *Beginning Mathematics*, Books 1-4 (Blackwell, 1968).
SERVAIS, W., article in *Mathematical Education in Science and Technology* (February, 1975).
SHAW, H. A. AND FUGE, K., *The Story of Mathematics* (Arnold, 1963).
SKEMP, R. R., *The Psychology of Learning Mathematics* (Penguin Books, 1971).
STERN, C., *Children Discover Arithmetic* (Harrap, 1953).
THORNDIKE, R. L. AND HAGEN, E. P., *Measurement and Evaluation in Psychology and Education* (Wiley, 1969).
WHEELER, D., article in *Trends in Education*, No. 23 (July 1971).
WORCESTERSHIRE EDUCATION COMMITTEE, *The Worcestershire Mathematics Progress Guide* (1970 and after).

Index

Accuracy 52, 53, 84, 91
Addition 65, 67, 70
Aims 105
Angles 89-91
Arbitrary units of measurement 81, 87
Area 24, 84, 86, 87
Area/perimeter relationship 86
Arrowgraph (arrowgram) 96
Associative law 74

Balance 81
Biggs, E. 35
Biggs, J. B. 104
Bilateral symmetry 41, 90, 95
Block graphs 96, 97
Broadcast programmes 118
Bruner, J. S. 103

Calendar 92
Capacity 80, 81, 87
Cardinal numbers 65
Cartesian co-ordinates 36, 56
Centicube apparatus 113, 114
Change: attributes 107; effects 122; vehicles of 116
Circle 24, 86, 91
Classifying 43, 44
Class teaching 119-21
Column graphs 96
Compass bearings 90
Complementary addition 72
Conic sections 95
Conservation of area 24, 85
Conservation of capacity 87
Conservation of measurement 81
Conservation of number 65
Co-ordinates 36
Cuboid 86
Cuisenaire apparatus 67, 68, 111, 113
Curriculum 32, 33

Decimal currency 92
Decimal fractions 76
Decimal notation 79, 93
Decomposition 73
Density 88
Derived units of measurement 80, 84
Descartes 36
Diagnosis 127
Dienes, Z. P. 103
Dienes Logic Blocks 34, 44, 45, 46, 47, 49, 50, 51, 52
Dienes Multibase Arithmetic Blocks 111
Distributive law 74
Division 71, 72, 74
Division by decimal fractions 79, 80
Division by fractions 78-80

Ellipse 56
Equal addition 72
Equalities 99
Equivalence of fractions 76, 77
Estimation 53, 84, 91
Euler, L. 56, 57

Fibonacci series 41
Flowchart 33
Formulae for area and volume 86
Fractions 75-80
Fundamental units of measurement 80, 84

Games (*see also* Dienes Logic Blocks) 115
Generalisation 55
Graphical representation 36, 37, 39, 95-100
Grouping (division) 71, 72
Group work 121, 122

Hexagon 26

Index

Histogram 99
History of mathematics 33

Identity element 78
Indices 75
Individual work 122
Inequalities 99, 100
Innovation 102
Instruments, use of 84, 91, 92
Inverse fractions 78

Length 80, 84, 85
Linear graphs 97
Linear programming 100
Logical thinking 45, 45–52
Long division 75
Long multiplication 75

Mass/weight relationship 82, 83
Measurement relationships 28, 44, 81, 88, 89
Measurement units 28, 35, 44, 80, 83, 84, 85, 86
Memory training 21, 23
Metrication 35, 82
Modular arithmetic 75
Money 92–3
Multiplication 18, 71, 73, 109
Multiplication tables 18, 38, 39, 73

Natural phenomena 55
Negative numbers 75
Nets of polyhedra 95
Networks 56
Number 17, 33, 65–80
Number bases 33, 75
Number classification 43
Number names 65
Number operations 18
Number pattern 27, 28, 38, 40
Numerals 65
Numerical competence 17

One-to-one correspondence 65, 96
Operations on decimal fractions 79, 80
Operations on fractions 76–9
Operations on money 93
Operations on numbers 65, 70–80
Ordered pairs 98
Ordinal numbers 65
Organisation 119–21

Parabola 42

Parallelogram 24, 86
Parents 129
Pendulum 29
Pentagon 25
Percentages 80
Perimeter 87
Pi 39, 95
Piaget, J. 103
Pictogram 96
Pie chart 99
Place value 66, 71, 74
Platonic solids 57
Polya, G. 54
Positive numbers 75
Practical skills 25
Pre-number stage 65
Prism 86
Probability 45
Problem solving 18, 22, 43, 54
Properties of shapes 94, 95
Protractor 91
Pyramids 54
Pythagoras 29, 95

Rag-bag of techniques 21
Ratio 80
Ready reckoner 93
Recording 68, 95
Record keeping 123, 127–9
Recurring decimals 80
Reptiles 32
Research 103
Resources 108, 117
Right angle 89, 90
Rigidity of shapes 94
Rotational symmetry 41, 90, 95
Rote learning 129

Samenesses 27, 49, 50
Scale 80, 97
Scatter diagram 99
Schemes of work used (list) 63, 64
Sealey Infants Mathematics Set 111, 113
Sealey, L. G. W. 43
Sealey textbooks 108
Sets 26, 57–60
Shapes 29–32, 93–5
Sharing (division) 71, 72
Short-cut methods 55
Similar shapes 87, 95
Skemp, R. R. 104
Sorting 44, 65
Speed 84

Spirals 41, 42
Standard units of measurement 82, 86
Stern apparatus 111, 113
Stern, C. 67, 111
Structural apparatus 105, 111–15
Subtraction 65, 68, 70, 72, 112
Surface area/volume relationship 126
Syllabus 61–63
Symbolism 36, 65, 68, 70
Symmetry 90, 94

Tangram 48
Television programmes 118
Telling the time 92
Templates 94
Tessellations 30, 31, 90, 94
Testing 127, 132–5
Textbooks 108–11, 124
Time 91–92
Transformations 91, 95
Trapezium 24, 86
Triangle 86, 91
Triangular numbers 27, 40
Twenty-four hour clock 92

Unifix apparatus 111, 113

Vectors 91, 95
Venn diagrams 43, 44
Vertices 56
Vocabulary 81, 92, 93
Volume 29, 81, 84–9

Weightlessness 33
Weight (mass) 80

Zero 65